Surviving Armageddon

Surviving
Armageddon
Solutions for a Threatened Planet

Bill McGuire

OXFORD
UNIVERSITY PRESS

OXFORD
UNIVERSITY PRESS

Great Clarendon Street, Oxford OX2 6DP

Oxford University Press is a department of the University of Oxford.
It furthers the University's objective of excellence in research, scholarship,
and education by publishing worldwide in

Oxford New York

Auckland Cape Town Dar es Salaam Hong Kong Karachi
Kuala Lumpur Madrid Melbourne Mexico City Nairobi
New Delhi Shanghai Taipei Toronto

With offices in

Argentina Austria Brazil Chile Czech Republic France Greece
Guatemala Hungary Italy Japan Poland Portugal Singapore
South Korea Switzerland Thailand Turkey Ukraine Vietnam

Oxford is a registered trade mark of Oxford University Press
in the UK and in certain other countries

Published in the United States
by Oxford University Press Inc., New York

British Library Cataloguing in Publication Data

Data available

Library of Congress Cataloging in Publication Data

Data available

ISBN 0–19–280571–1

1 3 5 7 9 10 8 6 4 2

Typeset by RefineCatch Limited, Bungay, Suffolk
Printed in Great Britain by
Biddles Ltd., King's Lynn, Norfolk

For Anna and Fraser McGuire

Contents

Illustrations

Foreword

Write it on your heart that every day is the best day in the year.
No man has learned anything rightly, until he know that
every day is Doomsday.

Ralph Waldo Emerson (1803–1882).

No matter how positive your outlook, the entrance
of the human race into the third millennium can
hardly be considered one of dazzling promise and
unbridled optimism. As far as perspectives on the coming
centuries are concerned, doom and despondency are without
doubt the new rock and roll. As the implications of climate
change have become ever more frighteningly apparent and a
global crusade against terrorism threatens to destabilize an
already creaking framework of nations, pundits and prophets
have fallen over themselves to inform us that it can't be long
before our cosy, comfortable world falls apart, or even disap-
pears up its own rogue physics experiment. I know, because I
have been just as guilty of promulgating gloom and despair
as the next eschatologist. In *A Guide to the End of the World:
Everything You Never Wanted to Know,* I considered—in what I
hope was a reasonably informed and balanced manner—
those global catastrophes that threaten our world and our

race: asteroid and comet impacts, volcanic super-eruptions, giant earthquakes, and mega-tsunami, the prospects for a new ice age and the coming hothouse Earth. As has always been the case for those who accept futurology's poisoned chalice of prediction, reaction to the book's publication covered viewpoints all and sundry. Wearing my writer's hat, as opposed to my volcanologist's helmet, I am always amused and sometimes confounded by how my literary scribblings incite such a wide-ranging panoply of responses, some almost embarrassingly flattering and supportive, others politically or ideologically motivated rantings that I always imagine to be accompanied by a ritual burning of the book, together with an effigy of its author. While considered by one reviewer 'an important book . . . that will cause readers to take a long-range view of life and history', and a work that is 'racy, pacy, opinionated, sassy and fun' by another (thank you Ted), *A Guide to the End of the World* appeared to trigger an apoplectic fit in the *Sunday Telegraph* science correspondent, who— writing in *New Scientist*— found it 'hard to believe anyone could have written a more hysterical account of our planet's future'. Clearly it is not possible to please everyone all of the time, but broadly speaking a thread of consistency did run through most critiques, best summarized, perhaps, by *Front Magazine*, whose reviewer suggested that 'if you like self mutilation, this book will make a humorous light read at bedtime. Otherwise you'll be shitting bricks for a week—and then worrying about methane levels in the lower atmosphere.'

I have little doubt that some readers were left running scared or, in the case of one senior citizen who regularly

1 Too many disaster books . . . Browsing the popular science section of a bookshop these days can be a particularly sobering and depressing experience.

contacts a colleague to check if it's safe to come out yet, barricaded in a basement flat with several hundred tins of corned beef for company. Without question, professional survivalists will have taken on board accounts of the threat posed by global catastrophes, nodded sagely at one another, and gone back to polishing their machine pistols with renewed vigour. This, however, was not the purpose of the book. Rather, its publication in 2002 constituted an awareness-raising exercise designed to drive home the point that our planet is a far more dangerous place to live than most of us appreciate. Its principal lesson taught that the period of relative cosmic, geological, and climatic calm during which modern society had developed and prospered could not last for ever. Throughout the 4.6 billion-year history of the Earth, our planet's crust had been pounded by asteroids and comets, rent by devastating earthquakes and volcanic super-eruptions, drowned by giant waves, and episodically buried beneath kilometre-thick ice sheets. While they were so infrequent that we had yet to see their like, such global geophysical events were not going to stop happening just because we had arrived on the scene. Furthermore, we were making prospects for a comfortable future far worse through triggering the most rapid period of climate change in recent Earth history.

Three years on, how do things look? Well, not much better, it must be said. The horrifying Indian Ocean tsunami has claimed more than a quarter of a million lives in over thirty countries – local and tourist alike – and provided a shocking and timely reminder that Nature's worst can affect the entire planet or a substantial portion thereof. The events of Boxing

Day 2004 have at last focused attention on the potentially enormous scale of future natural catastrophes. So far, however, there is little evidence to suggest that we will be better prepared next time. Furthermore, the events of 11 September 2001, and the nebulous war on terrorism that they spawned, have raised the prospect of endless civil strife in addition to the natural threats our ever more challenged society faces. The capability of our race to wipe itself out has attracted increasing re-examination: less this time with respect to the nuclear holocaust, but more in relation to rather more exotic terminations arising from new avenues in science and technology. Way out in front in the race for champion gloom-monger is former Astronomer Royal, Martin Rees, who—in his book *Our Final Century*—gives us just a 50:50 chance of surviving the next hundred years. No super-eruptions or asteroid collisions are implicated here, however; the end, according to Rees, is far more likely to be all our own work. Perhaps we will all disappear beneath a sea of grey nano-goo, surrender to the next bout of killer chicken virus, or disappear in a puff of space-time as an over-enthusiastic experimental physicist inadvertently triggers a phase transition in the state of the cosmic vacuum. Or maybe we will succumb to climate change—without question the most disturbing of all potential threats because its effects are already becoming apparent. Despite the protestations of a bunch of illiterati (at least when it comes to climate science) who continue—in the face of crushing evidence to the contrary—to peddle the message that contemporary global warming is a natural phenomenon and nothing to be concerned about, new research and observation has ensured that prospects for the impact of climate change over the next hundred years appear increasingly bleak.

The picture painted, then, continues to be far from a bright one—6 billion or so of us, shoehorned together on an overheating planet that is increasingly riven by pollution, natural catastrophes, man-made disasters, and civil strife. The prophets of doom are still out in force, either proselytizing on the imminence of Nature's revenge or portending the end of our race and our planet by our own hand. But can things really be so bad, and if they are, is there nothing we can do? With 4 million people killed by an estimated 50,000 natural disasters during the twentieth century, it seems we remain unable to cope with the common-or-garden threats of flood, storm, earthquake, and volcanic eruption. What then, could we possibly do if faced with the prospect of an asteroid impact or a volcanic explosion great enough to affect every-one on the planet? Well actually, quite a bit—provided we put our minds to it. Countless scientists and technologists, and others of a more esoteric bent, have been beavering away in recent years, bending their intellects towards protecting us from Nature's worst. Contemplation of the resulting ideas and proposals—some serious and considered, others wild and wacky—form the focus of this book.

Having once been told by a television director that I did not have the right face for a doom-monger—a little too round and jovial—I felt that only two choices were open to me: extensive and risky plastic surgery or the promotion of a more positive view of the future for our planet and our race. Hence this book is best considered as a sort of antidote to *A*

Guide to the End of the World. I should make it clear from the start, however, that I have not experienced a transformation like that of St Paul; no scales have fallen from my eyes and in no way do I now think the future will reveal itself to be all sweetness and light. As will become apparent, there are measures we may take to avoid, mitigate, or manage the worst effects of future global catastrophes, but that does not mean that we will necessarily take them. If the current ineffectiveness of the Kyoto Protocol aimed at reducing greenhouse gas emissions is anything to go by, there is a sufficient absence of political will even to address a catastrophe that is already upon us, let alone one that may lie thousands of years down the line. Furthermore, the chances are that many of the inventions or methodologies put forward as potential solutions to our problems may never be possible, while others carry such enormous risks that their use or implementation can never be sanctioned. Inevitably, science and technology are cast to play leading roles in tackling the worst Nature can throw at us in the future, and herein lies another problem. Gone is the post-World War II optimism, driven by the white heat of science, that saw the advent of nuclear power, man landing on the Moon, and the non-stick frying pan. Now this has been replaced by worries about the environment, the human condition, and the state of the world in which our children and their children will live. Science is no longer viewed by the majority as a cure for all ills; instead it is becoming increasingly regarded—true or not—as the source of many of the problems we face today. On the public's radar screen of science awareness, the conquest of space now barely registers—despite George W's election-year Martian crusade—while nuclear power as an energy miracle has just about dropped off the edge. Instead, shining bright and clear, bang in the screen's centre, are those issues that have the potential to impinge directly upon every inhabitant of the planet: human cloning, genetically modified organisms, climate change, threatening new diseases, and the rapidly expanding field of nanotechnology.

The judicious application of science and technology can help to solve some of the problems we have created for ourselves or that Nature forces us to address, but will a society increasingly mistrustful of scientists and technologists and their work permit this? How can society be persuaded, for example, that industrial-scale pumping of carbon dioxide into the deep ocean, as a means to reduce the concentration of the gas in the atmosphere, is a good and safe thing to do, when industrial technology has contributed in the first place to the bulk of a 30 per cent rise in greenhouse gas (GHG) emissions by releasing the gas into the atmosphere? How can the designers and builders of the world's nuclear arsenals make a convincing case for launching nuclear warheads over our heads and into space in an attempt to divert an asteroid that may or may not have our name on it? There is little doubt that techno-fixes to address future natural global threats will face considerable opposition. In some cases, this is no more and no less than they deserve. Swinging comets past the Earth in order to pull it into an orbit further from the Sun, thereby cooling it down, has recently been proposed by a NASA team. Clearly, such an outlandish scheme is going to struggle to find global acceptance ahead of simply living more sustainable and energy-efficient lives. One would hope that concerns over the *see-saw* effect—science and technology attempting to correct a problem they were responsible for, but making the situation worse, then trying another tack and making things worse still—are likely to prevent any such proposals being tried. Would you trust NASA scientists to determine correctly the new orbit needed for the Earth's temperature to be ameliorated, bearing in mind that in 1999 they lost a Mars probe because they failed to make a simple conversion from imperial units into metric ones?

Nevertheless, the application of science and technology is critical to reducing the impact of global natural catastrophes in the future. Without their twin benefits we will fail to have any real impact on climate change, nor will we be able to forecast a future volcanic super-eruption, or nudge off course an asteroid that is heading our way. Certainly science and technology together cannot be considered a panacea, nor will they provide a protective shield behind which our race and our planet can sleep soundly forever. In concert, however, they can present us with part of the solution to climate change and supply us with the means to detect potential global catastrophes far enough in advance, either to prevent them happening at all or, at the very least, to allow us some breathing space to prepare for the inevitable and maximize the chances of the fabric of our society surviving relatively unscathed. Crucially, a scientific and technological approach cannot be successful in isolation, but must be accompanied by other measures. In the case of climate change, these must involve modifying the way we live our lives, both as individuals and collectively. Similarly, our response to an asteroid impact or a super-eruption that we are unable to prevent is likely to entail drastic changes in the way our society currently operates, almost certainly involving changed priorities and a greater restriction on personal freedoms as we seek to recover and rebuild.

Rather than scaring the pants off you, I hope that this book will persuade you that although our future seems far from rosy, it is also far from desperate. In the opening chapter I take a look at the major geological, climatological, and cosmic phenomena that threaten our way of life, presenting the fruits of recent research, examining new ideas about scale and frequency, and evaluating the risk today. Having set the scene, the following three chapters address those options we have for avoiding, mitigating, or managing potentially catastrophic hazards that arise above us in space, in the crust beneath our feet, and all around us in the atmosphere. Finally, I take a little time to present a blueprint for a future in which a combination of science and technology, together with changes in the way we lead our lives, is able to ensure that the future of our race is one of bloom rather than doom. Ultimately, I hope that you will take away with you the message that all is not yet lost. Although our society will inevitably face knock backs, as a race we have the means within our grasp—through judicious application of our knowledge, accompanied by the adoption of a more considered life-style—to ensure that we, our planet, and all life upon it not only survive but also prosper.

1

The Heebie Gee-Gees:
Setting the Scene

There's no disaster that can't become a blessing
and no blessing that can't become a disaster.

Richard Bach: author, *Jonathan Livingston Seagull.*

On a glorious late spring morning in 1994, over a hundred geologists and volcanologists congregated in the august apartments of the Geological Society in Piccadilly's Burlington House, to discuss how and why volcanoes collapse: a catastrophic phenomenon most famously broadcast around the world during the climactic eruption of Mount St Helens in May 1980. The delegates were, as ever, a mixed bag: the inevitable gaggle of enthusiastic and mildly eccentric amateurs of independent means, one inseparable from his safari shorts, even in darkest winter, and a good number of scientists from less developed countries plagued by active volcanoes, including a husband and wife team from Russia, whose financial support from the UK's Royal Society I handed over rather surreptitiously in a brown paper bag. Most participants, however, were specialists from Europe and North America, including actor Jon Voight's brother Barry, an illustrious expert on volcano instability and landslides who is based at Penn State University, and a clutch of planetary scientists more concerned with collapsing volcanoes on Mars and Venus than on our own fair Earth.

As the conference organizer I had little time to listen to talks, instead attending to delegates' needs, checking return flight times, sorting out accommodation, or making threatening phone calls about stray luggage. As fortune would have it, however, I did have the opportunity during one of the refreshment breaks to chat to Spanish vulcanologist, Juan Carlos Carracedo—top man at the Volcanological Station of the Canary Islands, based in Tenerife. The Mount Teide volcano on Tenerife has been quiet since 1909 and concern about the contemporary volcanic threat there is, not surprisingly, low. But my interest was attracted by another volcano— the Cumbre Vieja—on the western Canary island of La Palma. This, it seemed, had erupted on a small scale in 1971, but far more interesting shenanigans had apparently been going on during a more violent eruption a quarter of a century earlier. Juan Carlos, who had an illustrated poster showing the geology of the volcano, pointed to a series of long fractures that had opened up on the volcano's flanks during an eruption in 1949. These accompanied strong earthquakes beneath the volcano's western slope and looked as if they might be evidence for the entire flank separating from the rest of the volcano and dropping 4 m towards the sea. In effect, a gigantic landslide poised, like a Damoclean sword, over the waters of the North Atlantic. Intrigued and enthused by what appeared to be the world's most recently activated giant landslide, collaboration swiftly ensued, and that autumn, with the help of financial support from the Spanish Research Council (CSIC), I led a team of colleagues and research students south to La Palma. Our objective was to establish a so-called ground deformation network: an array of survey benchmarks (actually just specially designed nails banged into stable rock outcrops) that would allow us to measure accurately and precisely the distances between them. This, we hoped, would tell us whether or not the landslide was still on the move.

Over the next five years, we returned to the island on a number of occasions to re-measure the network, first of all using a technique called *electronic distance measurement* and later the newer *Global Positioning System,* about which more in Chapter 3. By 1997 we had some idea of the landslide's behaviour and it did indeed seem to be still on the move, albeit very slowly. A research student, Jane Moss, who was studying the situation for her doctoral thesis, showed that benchmarks on the landslide over the period had all moved westward by about a centimetre relative to the rest of the Cumbre Vieja volcano. Although the coherence of the displacement pattern suggested that the movements were real, the small values were within instrumental error and could not therefore be taken as definitive. What was needed was further monitoring to build a picture of movement over a longer period of time. Alas, before more surveys could be undertaken and the situation clarified, events were to unfold that would thrust the anonymous little island into the global limelight and put any prospect of a return to La Palma on hold for the foreseeable future.

2 The Cumbre Vieja volcano occupies the southern third of the Canary Island of La Palma. During the 1949 eruption the volcano's western flank—a mass of rock that may be as large as the UK's Isle of Man—dropped 4 m. Sometime in the future it will crash into the ocean, threatening the entire North Atlantic margin with enormously destructive mega-tsunami.

In 1999 I published a book called *Apocalypse: A Natural History of Global Disasters*, in which I presented a fictional account of the future collapse of the western flank of the Cumbre Vieja and the devastating impact of the resulting giant waves—or mega-tsunami—on the cities bordering the Atlantic Basin. Having already been inspired to make an acclaimed programme about super-eruptions by another fictional account in the book, staff on BBC Television's flagship science series *Horizon* were on the lookout for more stories in the same vein. The Cumbre Vieja situation seemed perfectly to satisfy the viewing public's growing appetite for programmes about doom and disaster, and filming about the threat from La Palma started the following year, culminating in the broadcast in October 2000 of *Mega-tsunami: Wave of Destruction*. It would not be an exaggeration to say that all hell broke loose following transmission, as the world's media picked up on the story and requests for interviews and information piled up. A co-worker from University College London, Simon Day, who starred in the film, and I fielded questions and queries for weeks afterwards, some from the press, some from irate colleagues accusing us of scare-mongering, and others from weirdos and anoraks offering unlikely solutions to the problem or expressing the fear that terrorists could blackmail the US by threatening to use bombs to set off the collapse. All in all, the legacy of the *Horizon* programme was, first, to put the lovely but tiny island of La Palma well and truly on the global map, and second, to draw attention to how a relatively minor geological event in an insignificant archipelago could have the potential to trash half the planet. In essence, the film contributed towards a growing awareness of the fragility of our planet and the exposure of the human race to natural catastrophes on a scale that, due to their rarity, are unprecedented in modern times. Such *Global Geophysical Events* (GGEs) or *Gee-gees*, although extremely infrequent, are now very much regarded as part of the compendium of hazards that we need to consider when addressing the vulnerability of our society to natural phenomena.

Gee-gees are best defined as natural catastrophes on a scale sufficient either to impinge physically upon the entire planet, or to cause regional devastation accompanied by knock-on effects severe enough to damage the social fabric and/or the economy of the whole world. Fortunately, there are very few species of gee-gee and, in addition to an ocean-wide giant tsunami, the only rapid-onset calamities we have to lose sleep over are a great earthquake striking a major industrial centre, a large asteroid or comet impact, and a volcanic super-eruption. Operating rather more slowly, but still with a speed too great for society and life to adapt adequately, is abrupt climate change, which is currently taking place in the form of planetary warming fostered by human activities. The broad attribute that all gee-gees share is that they constitute so-called high-magnitude, low-frequency events. In other words, while their effects can be utterly cataclysmic, they occur on timescales that are many orders of magnitude greater than a single human life span. Consideration of a 'doomsday timescale' shows that the most catastrophic events of all occur only every 100 million years or more, taking the form either of mass outpourings of lava, sufficient to swamp a continent, or collisions between the Earth and an asteroid or comet in the 10 km size range. Both are *extinction level events* that, time and again, have wiped out a substantial percentage of life on Earth. Coincidentally, examples of both phenomena took their toll on our planet around 65 million years ago, and evidence is mounting for a link between the two; but I'll say more about that later. Smaller impacts, by objects a kilometre or more across, are far more frequent, but are still estimated to happen only once every 100,000–600,000 years, depending on whose figures you accept. While insufficiently large to wipe out the human race, the onset of freezing conditions arising from pulverized rock in the atmosphere blocking incoming solar radiation is estimated to be capable of killing tens of millions people. The immensely long time spans between such events might well foster a feeling of relative security, were it not for the fact that comparably shattering terrestrial events occur far more frequently. Volcanic super-eruptions, capable of triggering a volcanic rather than a cosmic 'winter', seem to have happened—on average—every 50 millennia over the last 2 million years. Giant tsunami, such as that threatened by the future collapse of La Palma's Cumbre Vieja volcano, may happen somewhere in the planet's oceans every 10,000 years or more, either due to disintegrating volcanic islands or from enormous slumps of sediment from the continental margins.

Because gee-gees happen so infrequently, and because one has yet to disturb our modern world, they have tended in the past to attract little attention, both within the scientific community and elsewhere. This all changed, however, on 16 July 1994—rather appropriately the twenty-fifth anniversary of the launch of the *Apollo 11* moon landing mission—when the first of 21 fragments of disrupted comet, Shoemaker-Levy 9 (or simply SL9) plunged into planet Jupiter's gaseous envelope. The impact ejected a huge plume of gas and debris surrounded by expanding shock waves. Over the next week, the remaining 20 fragments, some as small as 50 m, others kilometres across, crashed into the Jovian atmosphere. Most spectacular of all, was the impact of 4 km wide *Fragment G* that generated a flash so bright that it blinded many telescopes in the infrared wave band, and which left a great, dark spot on the planet's exterior that was larger than the Earth. With pictures of the collisions taken by the Hubble Space Telescope and the *Galileo* space probe *en route* to the planet, flashed around the world as they happened, it was not long before questions began to be asked. If a comet could hit Jupiter, could the Earth also be vulnerable? If the Earth was hit, what would it mean for us? Such concerns rapidly led to an enormous growth in interest in objects in space that offered a potential threat to our planet. Fledgling sky searches designed to spot asteroids approaching Earth were given a boost and even the UK government commissioned a task force to examine the threat of an impact. More than this, however, the events on Jupiter contributed to a sea change in the way we view our planet's well-being and security. This made later revelations about the threat from volcanic super-eruptions and giant tsunami somehow easier to accept and embrace as rare but natural phenomena firmly rooted in the real world as opposed to being abstracted from the realm of science fantasy. In a way, these revelations touting catastrophe on a hitherto unprecedented scale have also helped people to appreciate that our planet really is vulnerable—not only to the forces of nature but also to the activities of man.

But to return to the La Palma saga, which continues to rumble on. A year after the *Horizon* broadcast, Simon Day of UCL and Steve Ward, of the University of California at Santa Cruz, triggered another frenzy of media interest when they published a paper that modelled the tsunami that would follow from a future collapse of the Cumbre Vieja volcano. The terrifying results show an initial bulge of water almost a kilometre high subsiding to a series of waves many tens of metres high that batter the neighbouring Canary Islands and northwest Africa. Between six and twelve hours later, the giant waves savage the north-east coast of Brazil, the Caribbean, and the entire eastern seaboard of the US. Even the *Sun*, the UK's notorious 'red-top' tabloid newspaper could not resist the story, providing its readers with a list entitled 'but here's why a tsunami's not too bad'. Witticisms—and I am still not sure this is the correct term for what follows—from the list include, at number 5, 'the widespread flooding will save England having to follow-on against whoever our dismal cricketers are playing at The Oval,' and at 8 the embarrassingly xenophobic factoid that 'it makes us further from France and all those overpaid, euro-loving bureaucrats in Brussels.' Much to the chagrin of *Sun* readers and all virulent anti-Europeans, I feel I ought to point out that when the waters recede the distance from Dover to Calais will, in fact, be pretty much unchanged. And what of the volcano itself?

Things have been very quiet over the last few years. Local government is adopting the ostrich posture, attempting to deny that there is a potential problem here at all, and doing nothing to discourage the construction of new holiday properties on the landslide itself. The Spanish government also appears to pay scant regard to the problem. At this point, I really ought to do my bit for the local tourist industry and reveal that La Palma is the most fantastically unspoilt place, with a gorgeous climate, tremendous scenery, and wonderful food and wine. Visit it before it visits you. Surveillance of the volcano remains at a very low level, with just three seismometers listening out for any new magma on the move, and no one monitoring the landslide itself. A far from ideal situation, I am sure you will agree, but is the risk of this cataclysmic event really that high? Should the inhabitants of La Palma spend every waking hour worrying that they will open their eyes one morning to find their beautiful island considerably reduced in size? Must the hundreds of millions of people living around the rim of the Atlantic Ocean persistently lose sleep through concern about inundation by giant waves? The simple answer is no. Although the periodic formation of a giant landslide is part of the normal life cycle of an ocean island volcano, it is an event that occurs very rarely indeed, perhaps only once every 100,000 years or more at each individual volcanic island.

On La Palma itself, much of the extinct Taburiente volcano, which forms the northern part of the island, slid beneath the waves around half a million years ago, leaving a gigantic amphitheatre 14 km across enclosed by towering cliffs up to 2 km high. The Canary Islands have also hosted other enormous landslides, at least a dozen in all over the last several million years, and some of the islands have actually been shaped by the process. The neighbouring island of El Hierro, for example, has been sculpted by three collapses, each of which has formed a great concave gulf, the most spectacular of which is El Golfo on the northern side of the island. This is a colossal depression 20 km across that opens to the sea and is surrounded, on the landward side, by immense cliffs up to 1 km high. An explanation for the structure can be found offshore, where imagery of the seabed reveals a landslide extending 80 km from the island and containing individual blocks of rock up to 1 km across. The timing of the collapse is not well constrained, but it is believed to have occurred between 90 and 130 thousand years ago. The volume of the landslide looks as if it was about 100 cubic kilometres, which is equivalent to extracting the whole of downtown London, to a depth of 1 km, and dumping it in the ocean.

If these enormous masses of rock slid slowly beneath the waves, at perhaps a few tens or even hundreds of metres a year, there would not be a problem. All right, anyone unlucky enough to have their home on the slide would eventually get wet feet, but the threat of mega-tsunami would be absent. While nobody has yet observed a giant volcanic landslide entering the sea, it seems clear, however, that such an event would be over in just a minute or two. Crucial evidence for this can be found on the much truncated island of El Hierro, where a smooth, exposed surface known as the San Andres Fault marks the back wall of an aborted giant landslide. Here, during prehistoric times, an enormous block of rock slid seawards for 300 m before juddering to a halt, and close observation of the slip surface reveals why. In places there is still preserved a thin layer of a substance called *pseudotachylite*, a glassy material formed when rock is melted by extreme pressures. Landslides move most effectively along surfaces that are lubricated with water, which is why—incidentally—so many occur after heavy rain. The San Andres slide, however, seems to have been so dry, and moved so rapidly, that the enormous pressures generated at the slide's base actually melted the rock. Eventually, with no water to facilitate continued movement, everything was brought to a halt before yet more of the island disappeared beneath the waves.

The aborted San Andres collapse has great significance as it shows that Canary Islands landslides move rapidly and as coherent blocks, supporting the idea that their entry into the ocean is sufficiently catastrophic to generate tsunami big enough to threaten the devastation of ocean basins. And this is not the only evidence; on the neighbouring islands of Gran Canaria and Fuerteventura can be seen incongruous deposits containing wave-worn cobbles and seashells—the sort of thing you can find on any beach. These, however, are not on the coast but resting up to 100 m above the current sea level. In the absence of convincing evidence that the islands have been uplifted, the only explanation for these deposits is that they were dredged from the shore by a giant tsunami and dumped high and dry far above the sea. In fact, the waves responsible might have been even higher than the current height of the deposits suggests. The shell material they contain appears to reflect colder water conditions than those that prevail today around the Canary Islands, so emplacement may well have occurred during the last Ice Age, when sea levels were even lower. So far, it has not been possible to link the stranded tsunami deposits on Gran Canaria and Fuerteventura to any particular collapse event in the archipelago, but on the other side of the Atlantic Basin further enigmatic features may provide such a link.

When the Cumbre Vieja's west flank finally crashes into the ocean, the aforementioned Ward and Day tsunami model predicts that the resulting waves will have a terrible impact in the Caribbean, so this seems like a good place to search for the effects of past collapses. With the land rising majestically to a full 63 m above sea level, the thousands of islands that make up the Bahamas would offer little resistance to a megatsunami hurtling towards them from the Canary Islands. It is also likely that they would preserve some sign of such a catastrophic event occurring in the past, and it seems that this may now have been found. On the Bahamian island of Eleuthera, boulders of coral limestone from the seabed—as big as houses and weighing thousands of tonnes—have been stranded 20 m above sea level and a good half-kilometre inland. How did they get there? The only reasonable explanation is that they were catapulted into position by extremely energetic waves. It has been proposed that unparalleled storm waves may have accomplished this, and certainly storms are capable of lifting rocks of this enormous size and dumping them onto beaches or cliffs. It is highly unlikely, however, that wind-driven waves could manage to transport such weights half a kilometre inland. On the other hand, giant tsunami would be eminently capable, as their much longer wavelengths (crest-to-crest distances) ensure that rather than crashing fleetingly onto a shoreline before rapidly dissipating, they flood inland as a wall of water of unprecedented power. Even a relatively commonplace tsunami generated by a submarine earthquake is capable of delivering an incredible 100,000 tonnes of water for every 1.5-metre stretch of coastline, so the shear energy available to a giant tsunami makes it easy to cope with shifting a house-sized boulder a few hundred metres. Great wedges of sand, several kilometres long, and squeezed through the gaps between the islands, may also be a legacy of great prehistoric tsunami. Encountered all along the archipelago, these sand wedges have again been linked to exceptional storm waves that might have accompanied climate deterioration around 120,000 years ago. But as for the boulders, another explanation is possible. At about the same time, the Canary Island of El Hierro was spontaneously shedding around 100 cubic kilometres of rock into the North Atlantic, leaving behind the El Golfo amphitheatre and sending a series of giant ripples hurtling westwards. Just eight hours later, the same waves would surge across the Bahamas *en route* to the North American coast.

I don't want to give the impression that the Canary Islands archipelago is a hotbed of unstable volcanoes and tsunami sources, unique in the world. Far from it: evidence for enormous landslides is encountered on or around many of the world's volcanic island chains, including the Hawaiian Islands, which are encircled by a vast apron of debris accumulated as a result of close to 70 distinct collapse events, each of which may have sent giant tsunami surging across the Pacific Basin. Vast embayments on the flanks, or landslide debris scattered across the adjacent sea floor, testify to many other landslides occurring around coastal and island volcanoes during prehistoric times, including at Etna and Stromboli in the Mediterranean, Tristan da Cunha and the Cape Verde Islands in the Atlantic, and the island of Réunion in the southern Indian Ocean. Furthermore, giant landslides in marine environments are not confined purely to volcanoes. In 1993, David Smith and colleagues from Coventry University identified a rather innocuous-looking sand layer buried deep within the peat of Sullom Voe in the Shetland Isles. The same layer is encountered across much of northeast Scotland, and marks a time—around 8,000 years ago—when tsunamis surged across the country with the power of an express train. Flint tools found in the layer close to Inverness suggest that the waves obliterated at least one Mesolithic community, the first tsunami victims that we know of. This time the source of the waves was not a disintegrating volcano, but a truly gargantuan slide of submarine sediment. As sea levels rose rapidly following the end of the last Ice Age, the margins of the continents became increasingly unstable. Fresh sediment washed down from land newly exposed by the retreating ice accumulated in ever greater volumes on the edges of continental shelves that were periodically rocked by large earthquakes as the crust at last divested itself of the great load of ice that had kept it pinned down for tens of thousands of years. Perhaps it was one of these quakes that triggered the collapse of the *Storegga slide*—3,500 cubic kilometres of sediment from the continental margin west of Norway, generating one of the largest known landslides on the planet. In fact the area of landslide debris on the floor of the Norwegian Basin is about the same size as Scotland. Evidence for a tsunami generated by the slide has now been found in Norway itself and in the Faeroe Islands, with estimates of the wave height as it struck land ranging up to 25 m. Given the scale of the event, it is almost certain that the waves flooded coasts much further afield, and probably affected most coastlines in the North Atlantic.

As previously mentioned, in the normal run of things, the formation of a giant tsunami is a very rare event, happening somewhere on the planet—on average—about every 10,000 years or so. A nagging suspicion is growing within gee-gee circles, however, that such phenomena may actually be clustered in time, occurring more frequently when the climate is undergoing rapid change. Even as I write this, a paper published in the prestigious journal *Geology* links submarine collapses such as Storegga to the swift changes in sea level that accompanied the advance and retreat of the glaciers during the last Ice Age. A connection has also been proposed between the timing of volcanic island collapses and the warmer, wetter intervals that punctuated the bitter cold of the last Ice Age. At a time of rapid climate change, when sea levels are once again on the rise and the climate is forecast to get wetter over the volcanic islands of Hawaii and the Canaries, such correlations are, to say the least, disconcerting.

It is hardly difficult to imagine how the simultaneous fall of the great cities of the Atlantic rim is likely to play havoc with the economic and social fabric of the global community, bringing a crash far worse than that of the 1920s, followed by a slow and tentative recovery requiring many decades. But what about the destruction of a single city—could that have the same impact? Does any urban centre on the face of the planet carry such global clout that bringing it down would have serious ramifications worldwide? Well, there is just one—the sprawling metropolis of Tokyo on the Japanese island of Honshu. Once already in the last hundred years, this huge expanse of densely packed humanity has succumbed to obliteration—a legacy of one of the titanic earthquakes that strike this tectonically unstable region with unnerving frequency. Now, once again, the 30 million or so people crammed into a crescent of land around the north end of Tokyo Bay face the prospect of annihilation.

September 1st is disaster prevention day in the Japanese capital, and for very good reason. In 1923, a few minutes before noon on the first day of autumn, the Earth's crust failed to the south of the city. Eighty kilometres away, beneath the waves in Sagami Bay, a major fault finally succumbed to enormous strains that had been accumulating since the last great earthquake in 1703, and began to tear itself apart. Even as the capital settled itself down to lunch in packed tea-rooms, cafés, and beer halls, powerful pulses of ground-shaking were hurtling northwards, battering first the city of Yokohama and crashing into Tokyo just 40 seconds later. The flamboyantly named British businessman—Otis Manchester Poole—was unlucky enough to be working in his Yokohama office that fateful morning, and described the impact of the quake in his book, *The Death of Old Yokohama*, thus:

> I had scarcely returned to my desk when, without warning, came the first rumbling jar of an earthquake, a sickening sway, the vicious grinding of timbers and, in a few seconds, a crescendo of turmoil as the floor began to heave and the building to lurch drunkenly . . . The ground could scarcely be said to shake; it heaved, tossed and leapt under one. The walls bulged as if made of cardboard and the din became awful . . . For perhaps half a minute the fabric of our surroundings held; then came disintegration. Slabs of plaster left the ceilings and fell about our ears, filling the air with a blinding, smothering fog of dust. Walls bulged, spread and sagged, pictures danced on their wires, flew out and crashed to splinters.

Disagreement exists about the duration of the shaking, but reports talk of between four and ten minutes of continuous vibration, with near constant motion persisting for perhaps two and a half hours. As with most major earthquakes, however, it is likely that the severest shaking that caused most destruction lasted for just a few tens of seconds. In Yokohama and Tokyo this was all the time it took to reduce the twin cities to rubble. In total some 360,000 buildings were destroyed, including 20,000 factories, 1,500 schools, and the great Imperial University Library, at the time one of the world's foremost repositories of works of art and old books. In Tokyo, 71 per cent of the population lost their homes, with this figure rising to over 85 per cent in Yokohama. Out of a population of 11.7 million, 104,000 were killed and a further 52,000 injured, with 3.2 million people left homeless. The quake registered 8.3 on the Richter Scale and 7.8 on the national scale used by the Japanese Meteorological Agency—a great earthquake by any standard. By far the worst destruction occurred, however, not as a result of the severe ground-shaking, but due to the terrible conflagrations that raged for more than two days afterwards. In homes across the region, the shaking had overturned countless *hibachis*, the traditional open charcoal burners popular for preparing meals. Within minutes the toppled stoves had started thousands of small fires that rapidly spread and merged to form larger blazes. As misfortune would have it, the weather was very hot with gusty winds, ideal for the growth and spread of fire. In the older, crowded parts of the cities, especially where wooden buildings predominated, blazes were rapidly transformed, first into infernos and then into giant walls of fire that marched unstoppably across the cities, consuming all in their paths. With water mains cut by the quake, and augmented by fractured natural-gas pipes that spouted fire, by exploding munitions and stored fuel and chemicals, the conflagrations achieved such a scale that the fire services could do nothing to prevent their progress. While those incarcerated in the rubble of collapsed buildings were burned alive, tens of thousands of battered and stunned survivors fled before the flames, seeking safety in the sea, in rivers and open spaces. But there was to be no refuge from the fire storms. Thousands were immolated by burning oil pouring into the sea from ruptured storage tanks, while others were slowly boiled in river water heated by the flames. The greatest tragedy occurred within the confines of a vacant plot of land about the size of a soccer pitch, known as the Honjo Military Clothing Depot. By mid-afternoon, over 40,000 terrified men, women, and children were crammed into the space, praying for deliverance from the encircling barrier of fire. Inexorably the temperatures rose to over 80 degrees C, triggering strong winds that created tornadoes of fire and sucked the firestorms together, overwhelming those trapped in the depot. All but a few hundred appallingly burned survivors were roasted alive; the clothing, bedding, and furniture that they had rescued from their flattened homes provided a perfect source of fuel for the greedy flames. Great heaps of charred bodies greeted the first rescue workers to come across the scene. Some people died standing, held upright by the press of bodies as the crowd shrank back from the encroaching flames.

The Great Kanto earthquake, as it is now known, provides a fine example of how a single natural catastrophe—if sufficiently large—can have repercussions far beyond the area immediately affected. The worst natural disaster in the country is estimated to have cost around 50 billion US dollars at today's prices, and proved an unsustainable drain on the national economy. Together, the quake and the global economic crash that followed six years later, triggered economic collapse and plunged the country into deep depression. In a mirror image of Germany's Weimar Republic, the resulting climate of despair and misery fed the growth of fascist doctrine that promised to make Japan great again, leading ultimately to the rise of the military, an unquenchable thirst for empire, and total war.

Eighty years on the cities of Yokohama and Tokyo are rebuilt and thriving. The two are now part of a much larger urban entity that goes by the name of the Greater Tokyo Metropolitan Region, a gigantic agglomeration of 33 million people—some 26 per cent of the nation's population—and the largest urban concentration on the planet. The terrible events of 1923 are far from forgotten. They occupy hearts and minds at the beginning of autumn every year, when natural disasters and especially earthquakes are very much the order of the day. Talks, exhibitions, and drills draw attention to the ever-present seismic threat, but are they serving any useful purpose? Can they truly help to prevent another catastrophe on the scale of 1923, or is it inevitable that the Japanese capital must once again succumb to Natures seismic shock troops? Several forecasts do indeed warn of a coming catastrophic quake in the region, the effects of which will once again reverberate across the world. Many of the most modern buildings in Tokyo and Yokohama are constructed according to stringent building codes that ought to ensure their survival during the next great quake. Countless others, however, are not. The Tokyo metropolis alone is estimated to hold a million wooden buildings, which will provide excellent fuel for the fires that will inevitably arise after the ground stops shaking. Many more buildings are constructed on nearly 400 square kilometres of land reclaimed from the waters of Tokyo Bay since the Great Kanto earthquake, land that will rapidly liquefy as seismic waves pass through it, causing buildings to founder and collapse. With concern growing over the future fate of the Japanese capital, the United States Geological Survey and one of the world's biggest reinsurance companies, Swiss Re, have come together to undertake a new assessment of the seismic threat. Prospects for major loss of life and destruction were increased following the 1995 Kobe earthquake 400 km south-west of Tokyo, which took over 5,000 lives, destroyed or damaged more than 140,000 buildings, and cost around 100 billion US dollars, making it the most expensive natural catastrophe of all time. The devastation wrought by the Kobe quake, which registered only 7.2 on the Richter Scale and lasted for just 20 seconds, and the apparent inability of the authorities to cope with the aftermath, bode far from well for the fate of the capital after the next major earthquake, and predictions of its likely impact dwarf that of Kobe. Despite improved building construction and a better understanding of the risk, a threefold rise in the population of the region is predicted to see up to 60,000 lives lost when the next major quake strikes, with the cost of the event perhaps totalling a staggering 4.3 trillion US dollars— up to 43 times more than Kobe. After over a decade of stagnation and the accumulation of a gigantic government debt one and a half times the country's GDP, serious concerns are already being voiced about the possibility of a future collapse of the Japanese financial system and a resulting global economic turmoil. A devastating earthquake striking at the heart of an already fragile economy could just be the metaphorical last straw. But more on this in Chapter 3.

Some natural phenomena do not need to rely on bringing chaos to the global economy in order to impinge upon every man, woman, and child on the planet. Their direct physical effects on our world are simply so enormous that there is no escaping them. Collisions with asteroids and comets fall into this category, and so do the greatest volcanic eruptions. To the west of the Indonesian island of Sumatra the ocean floor is cut by the 2,600-km long Java Trench, a slash across the seabed over 7 km deep that marks the contact between two of the rocky tectonic plates that make up Earth's 100 km or so thick, rigid, outer carapace. Beneath Sumatra the north-eastward-creeping Indo-Australian plate plunges back into the planet's baking interior, partially remelting to generate a supply of runny *basalt* magma that begins to wend its way back towards the surface. *En route*, however, it encounters the *granite* rock that makes up the crust underlying Sumatra, which it melts in turn to form particularly sticky magma, known as *rhyolite*, that is richly charged with gas. As a result of a relatively low density, enormous masses of this sticky rhyolitic liquid push upwards like the slowly rising globules in a lava lamp, most cooling and coming to rest several kilometres before they reach the surface. Very occasionally, however, a rising body of molten granite finds and exploits a weakness in the overlying crust. When this happens, the result is that unparalleled explosion of magma and gas—a volcanic super-eruption.

Northern Sumatra hosts just such a weak point, adjacent to the so-called Sumatra Fault Zone, which has on three occasions provided a conduit for rising magma blasting its way towards the surface. The legacy is the spectacular Lake Toba, at 100 km across the greatest volcanic crater on the planet; now water-filled and dormant, but still offering the near certainty of more massive eruptions to come. The crater is the combined product of three huge eruptions during the last million years, the oldest occurring as far back as 788,000 years ago, but the youngest blanketing the region with ash and debris only 73,500 years ago—just yesterday on the 4.6 billion-year geological timescale. As far as we know, this was the last time that the Earth's surface was punctured by a volcanic super-eruption, and with a couple of these events happening every hundred millennia or so over the last two million years, you could say that we are well overdue for another. Then again, the Earth does not operate like clockwork, but often more like buses in towns; things are quiet for ages and then two or three dramatic events happen in close order. The year 1991, for example, saw two major volcanic eruptions—at Pinatubo in the Philippines and Cerro Hudson in Chile—within months of one another when, by rights, eruptions on this scale should happen only every century or so. Likewise, the next super-eruption might inflict itself upon us in 50 years' time, or we might have another 50,000 years to wait.

It is incredibly difficult to get across some idea of the scale of the biggest eruptions on record. Most people's idea of a huge eruption is that which obliterated the upper half-kilometre of Washington State's Mount St Helens volcano in May 1980: the most filmed and televised of all eruptions, as you might expect, given its location in the world's most media-mad country. On the volcanic equivalent of the Richter Scale, however, the Mount St Helens blast registered as something that was very much run of the mill. The rather cumbersomely termed *Volcanic Explosivity Index* was devised in 1982 by volcanologists Chris Newhall and Steve Self, in order to provide some measure of the relative sizes of volcanic eruptions and their violence. The scale starts at zero for so-called *effusive* eruptions that involve the quiet extrusion of fluid lavas, with higher points on the scale matched to ever more explosive events that eject increasing volumes of ash and debris. Like its seismic equivalent, the VEI is logarithmic, which means that each point on the scale represents an eruption ten times larger than the one immediately below. The 1980 eruption of Mount St Helens, which scored a 5, was therefore ten times bigger than the VEI 4 eruption that buried the town of Rabaul on the Pacific island of New Britain in 1994, but ten times smaller than the explosion that ripped apart the Indonesian island of Krakatoa in 1883. Where, then, do super-eruptions fit into the scheme? Exactly where you would expect to find them—at the top—or at least what is probably the top. The Index is actually open-ended, again like the Richter Scale, but super-eruptions score an 8, making them at least a thousand times more violent than the Mount St Helens eruption and a hundred times bigger than Krakatoa. There may, in the distant past, have been volcanic blasts large enough to register a 9 on the VEI, but so far at least we have found no convincing evidence for this. Perhaps there is only so much volcanic energy the Earth's crust can contain before breaking, and maybe that limit is defined by a VEI 8 event.

In addition to providing a guide to the violence of an eruption, the VEI also takes account of the volume of material ejected. Events on the Mount St Helens scale involve the release of a measly cubic kilometre of magma or thereabouts; still enough, however, to bury the whole of Greater London beneath an ash layer a metre thick. Krakatoa-sized events, on the other hand, expel around 10 cubic km of debris, sufficient to blanket the Home Counties under a similar thickness. Impressive as these statistics sound, they pale into insignificance in comparison to the volume of material blasted out by Toba's last eruption, which—even using the lowest estimate of 2,800 cubic kilometres—would be quite enough to entomb the *entire* UK beneath an ash layer 4 m thick. Then again, every statistic that relates to a super-eruption is equally staggering. The violent mixture of magma and gas is released at a rate 120 times that of the flow of water over the Victoria Falls, and blasted 40–50 km to the edge of space, before spreading far and wide through the upper atmosphere. No one knows how long such an eruption would last, but it could be a week or two, going full pelt—many times longer than the most violent volcanic events of historical times.

Without doubt a super-eruption is capable of devastating an entire region, if not a continent. Some 640,000 years ago at Yellowstone in Wyoming, for example, a huge blast pumped out ash that fell across 16 states and as far afield as

El Paso (Texas) and Los Angeles (California). Hurricane-force blasts of hot ash and incandescent gas, termed *pyroclastic flows*, scoured an area equivalent to the size of an English county or a small American state. Going back much further to the Ordovician period some 450 million years ago, an eruption occurred—possibly in what is now the state of Ohio—that was so cataclysmic that deposits of its ash are encountered in rocks across millions of square kilometres of the eastern United States, Europe, and Russia.

Notwithstanding their capability for regional or even continental devastation, however, it is their potential for drastically affecting the global climate that qualifies super-eruptions for gee-gee status. In a seminal paper published in the journal *Nature* in 1992, Mike Rampino, of New York University, and Steve Self, a Brit then at the University of Hawaii and now at the Open University, proposed that the Toba eruption 73,500 years ago lofted sufficient ash and gas into the atmosphere to trigger severe global cooling—a phenomenon that has since come to be known as *Volcanic Winter*. In terms of their impact on climate, it is not so much the ash ejected by super-eruptions that is the problem; this settles out fairly rapidly. More important is the volume of sulphur-rich gases—mainly sulphur dioxide—that the eruption injects into the atmosphere. Here, the gases readily combine with water vapour to form tiny sulphuric acid droplets known as *aerosols*, which can spread rapidly across the planet, forming a thin but effective veil capable of significantly reducing the amount of solar radiation reaching the Earth's surface. Rampino and Self calculated that the Toba eruption yielded between

1,000 million and 5,000 million tonnes of sulphuric acid, sufficient to bring worldwide gloom and dramatic cooling lasting several years. In the tropics, temperatures may have plunged by up to 15 degrees C, with falls of 3–5 degrees C occurring across the planet. It may or may not be coincidence that the Toba explosion took place at a time when the Earth was already cooling, with glaciers on the move and sea level falling rapidly as more and more ice was locked up in growing ice sheets. Rampino and Self have suggested, however, that the Toba cooling may have provided that last little kick needed to accelerate our planet into full Ice Age conditions. And the impact of the eruption may not end there. Stanley Ambrose, an anthropologist at the University of Illinois, has seized upon the Toba aftermath as a possible explanation for a human population crash that seems to have occurred around about this time, with the number of individuals perhaps reduced to a few thousand for several millennia.

It is a sobering thought that a single volcanic eruption may have placed our ancestors in a position as precarious as that of today's Sumatran tigers or white rhinos, but looking back further in Earth history, volcanic eruptions are charged with an even greater impact on life. The term super-eruption is reserved strictly for cataclysmically explosive events. There are also, however, volcanic events that erupt unimaginably more magma, but which do so calmly and quietly. These so-called *flood basalt* eruptions are located above hot spots or *plumes* in the Earth's mantle, and involve the effusion of gigantic volumes of lava, with little—if any—accompanying ash. Because of their low viscosities, the lavas spread rapidly to cover enormous areas that are difficult to miss. Huge outpourings are encountered in the north-west United States, India, South Africa, and even in northern Scotland. The greatest flood basalt eruption so far recognized dwarfs them all. At the end of the Permian period, some 250 million years ago, northern Siberia hosted a veritable ocean of lava, known as the *Siberian Traps*, that spread across more than 25 million square kilometres—three times the land area of Australia.

A debate still rages about the role, if any, these huge outpourings have had in triggering *mass extinctions*—dramatic die-backs in Earth history that saw the sudden and simultaneous disappearance of many species. In the Permian period, prior to the Siberian outburst, for example, the Earth teemed with life. Such a fecund world was soon to be transformed into a wasteland, with only 5 per cent of Permian species making it through to the following Triassic period. Similarly, another mass extinction that occurred around 65 million years ago—this time with only 60 per cent of species vanishing—has been attributed to the great lava outpourings in north-west India that are known as the *Deccan Traps*. The precise mechanism whereby these lava floods are able to wipe out a significant proportion of our planet's life forms remains to be established. One favourite, however, is that enormous quantities of carbon dioxide pumped out during the eruptions—which probably lasted for hundreds of thousand of years—may have led to severe global warming and the annihilation of species that were unable to adapt rapidly enough to the new conditions. Surely a salutary warning at a

3 A volcanic caldera, such as Crater Lake in Oregon, is formed when a major explosive eruption evacuates the magma reservoir and the crust collapses into the void created. Yellowstone caldera in Wyoming is 80 km across, while that formed 73,500 years ago by the Toba super-eruption is 100 km in length.

time when greenhouse gas (GHG) concentrations in the atmosphere have risen by a third in the last 200 years and continue to rise at an ever accelerating rate.

The evidence for a link between mass extinctions in Earth history and episodes of flood basalt production is strong, but there is equally robust support for such extinctions arising due to collisions between our planet and wandering asteroids and comets. This is the essence of one of today's most vigorous debates in the Earth sciences: do flood basalts cause mass extinctions or are they the result of large impact events? Even more intriguing, do the impacts themselves trigger flood basalt eruptions, with the severe environmental effects of the two conspiring to make conditions increasingly difficult for many species? This often vociferous, and sometimes rather belligerent, debate is currently unresolved and is likely to run for some time yet.

Mention of impact events brings me rather neatly to the fourth gee-gee, a collision between our planet and one of the countless chunks of rock that are constantly hurtling across our solar system. As declared earlier, the impact menace was the first geophysical phenomenon to be recognized as having the potential to cause a global predicament. Since the Comet Shoemaker-Levy impacts on Jupiter in 1994 there has been considerable progress in addressing the issue. The success in spotting new objects with the potential to threaten the planet can be measured by the billboard headlines that scream at us, seemingly every couple of months: 'Asteroid to hit Earth' or something similar. Inevitably, after fumbling for change and grabbing a copy, we find that the chances of a collision are so small that we are far more likely to die as a result of an argument with our trousers. Apparently, over 5,000 people were injured in such altercations last year, so the chances are that this item of apparel will take a life soon. The threat of asteroid collision is nevertheless a real one, and eagle-eyed spotters of asteroids already have a possible candidate for collision on their books. The bad news is that the 1 km wide asteroid 1950AD has as high as a 1 in 300 chance of striking the Earth. The good news is that this might happen on 16 March 2880, so we do have a little time to think about how to tackle the situation. First spotted in that year, 1950AD was observed by astronomers for a little over two weeks and then 'lost' for half a century, only to be rediscovered on New Year's Eve 2000, no doubt by an observer who was not the partying type. The precise orbital characteristics of many newly discovered objects are often poorly constrained at first and the chances of collision of apparently threatening asteroids are usually revised downwards as their future paths become better known. Because the orbit of 1950AD has been observed across a 50-year period, however, it has been very well determined, so the relatively high chance of collision is real and unlikely to change that much as we approach the big day.

It is rather comforting to know that despite increasing numbers of methodical sky surveys, only one object has been spotted that presents any sort of problem at all. Nevertheless, experts on the threat of impacts have felt the need to devise a scale designed to inform the public of the level of risk presented by any object heading our way. The *Torino Scale* is a revised version of an asteroid and comet hazard index originally devised by Richard Binzel at MIT and adopted at a conference held, not surprisingly, in Turin. This 11-point index is designed to categorize the threat posed by newly discovered asteroids and comets whose initial orbital characteristics suggest they might approach the Earth at some time in the future. Virtually all new objects immediately score a zero—'events having no likely consequences'. In a very small minority of cases, however, the orbital data are sufficiently poor that the risk of a very close approach, or even a collision, cannot be ruled out. These objects may initially score higher on the scale, although rarely more than 1—'events meriting careful monitoring'. Once their orbits have been better determined, however, this has always dropped to 0, except, that is, for 1950AD, which is the first object to have its category 1 status confirmed. Categories 8, 9, and 10, defined as 'certain collisions', are the ones we all want to avoid, especially 10, which is reserved exclusively for collisions capable of causing global climatic catastrophes. With such incidents estimated to occur every 100,000–600,000 years, the chance of our experiencing an impact event on this scale in the course of the next hundred years is extremely small. Nevertheless, the threat is a genuine one and our planet will continue to be periodically, if infrequently, bombarded by objects meriting gee-gee status as far ahead as we can see.

While the appearance of a comet—or 'hairy star' as they were termed in medieval times—has always been viewed as a portent of doom, only very recently has any serious consideration been given to objects from space actually landing on

Earth. Even 250 years ago, the idea that some rocks found on the surface came from space was thought quite mad, and nothing could dissuade the scientists of the time from this viewpoint. As recently as 1768, a still smoking meteorite on the ground was decreed to be a rock that had been struck by lightning, and not until later in the eighteenth century did it become accepted that some rocks did indeed come from beyond the atmosphere. Space rocks continue to rain down upon the planet, and something like 3,000 tonnes of dust and small pieces of stone and iron are intercepted by the Earth every day. The vast majority burn up in the atmosphere, ending their lives in a blaze of glory as meteors or 'shooting stars'. Only about 100 bits of debris are large enough to survive a passage through the atmosphere and reach the surface as *meteorites* every year, which makes the story of Roy Fausset a particularly extraordinary one. On 23 September 2003, Roy returned to his New Orleans home to find that a direct hit from a meteorite had completely wrecked his newly renovated bathroom. It is not the bathroom-smashers, however, that are the real problem, but the very much larger objects capable of obliterating a city, a country, or a substantial part of the planet. Broadly speaking these take two forms: the asteroids, which are rocky bodies with near circular orbits about the Sun, and the comets, typically larger and faster masses of rock and ice whose highly elliptical orbits can take them far beyond the edge of our solar system on journeys lasting many thousands of years.

Most asteroids travel around the Sun in the *asteroid belt* between the orbits of Mars and Jupiter and rarely threaten our planet. Others, however, follow paths that either approach or intersect the orbit of the Earth, and these are the ones that need watching. In the world of impact hazard science the acronym is king, so these Earth-threatening rocks are known as NEAs or *Near Earth Asteroids*. They should not be confused with NEOs, *Near Earth Objects*, which also include local comets that approach our planet's neighbourhood. Nor should they be confused with PHAs, *Potentially Hazardous Asteroids*, or PHOs, *Potentially Hazardous Objects*, which incorporate, respectively, asteroids and asteroids plus comets, whose orbits bring them close enough to menace the Earth with collision. Asteroids make up the overwhelming majority of NEOs, and as of January 2005, 3,167 NEAs had been spotted, 757 of which are rocks with diameters of a kilometre or more. The number of PHAs is currently 669, with only 1950AD having any realistic chance of collision in the foreseeable future. Local comets, such as the famous Halley, follow rather eccentric paths that carry them from the inner solar system to its margin and then back again. Unlike those whose journeys take them perhaps halfway to the nearest star, they take just a few decades to make a single orbit—76 years in Halley's case—and none of those known currently poses an immediate threat of collision.

While it is good to know that thousands of potentially threatening objects are being tracked every day, it would be even nicer to know that we had spotted all of them and could sleep safely in our beds without the worry of being awakened by a collision with a rogue asteroid that ruins more than just our bathroom. Progress is being made in this respect, but

Empire State Building

4 Eros is a 24 km wide Near Earth Asteroid, which fortunately is not in danger of striking the Earth for as far into the future as we can predict its orbit. Detailed images of Eros were returned to Earth from the *NEAR-Shoemaker* probe, which—in February 2001—eventually touched down on the asteroid's surface.

there remains some way to go. The main problem is that it is difficult to pin down just how many threatening objects are out there, and new estimates of numbers appear every few months. Up to 20 million lumps of rock over 10 m across could be hurtling across the Earth's path during its orbit around the Sun, perhaps 100,000 of which could be 100 m or more in diameter—large enough to wipe out London, Paris, or New York. Around 20,000 are estimated to be around half a kilometre across, sufficiently large, should they strike the Earth, to obliterate a small country or generate a devastating tsunami if they hit an ocean. Of most concern, however, are those objects large enough to qualify unequivocally as potential gee-gees. The size of an object capable of physically disturbing the entire planet and everyone on it remains a matter of conjecture and enthusiastic debate, and I will return to it in the next chapter, but there is broad agreement that a collision with anything 1 km in diameter or larger is likely to have global ramifications. Predictions of the number of Near Earth Asteroids in this size range vary, but somewhere around a thousand seems an acceptable number. With more than 750 of these big bruisers already recognized, this leaves about a third of the total still to find. NASA hopes to have identified 90 per cent by 2008, but finding the balance may take substantially longer.

Later in the book I will examine in some detail the multitudinous plans put forward to see off objects on a collision course with the Earth. But if a large asteroid should get past our defences or appear on the horizon before we have been able to make any preparations, exactly what damage would it cause? A collision with a 1-km object would generate an explosion comparable to 100,000 *million* tonnes of TNT (this is equivalent to nearly two-thirds of a million Hiroshima bombs), which would excavate a crater about 25 km across and a fireball that would ignite everything within a radius of 150 km. Such a strike centred on London would instantly obliterate most of England, killing tens of millions of people. In the longer term, dust and debris lofted into the atmosphere would be expected to cause significant global cooling lasting for several years. Average death tolls for such an event have been estimated at around 60 million. An asteroid twice this size would have more serious local and regional consequences commensurate with its increased diameter, but global consequences would be disproportionately far harsher. A true cosmic equivalent of volcanic winter is predicted, with darkened skies and severe cooling leading to the collapse of agriculture and the resultant deaths of perhaps a quarter of the human population. Anything larger than 2 km and we could pretty much say goodbye as a viable species and join the ranks of the dinosaurs and millions of other species wiped out in the great mass extinctions of the past. Collision with a 10-km object would excavate a crater approaching 200 km across, create a fireball that ignites everything within a 2,000-km radius, and scatter molten debris across the entire planet. The apparent ramifications of an impact of this size that struck just off the coast of Mexico 65 million years ago included super-hurricanes, gigantic tsunami, global fires, perpetual darkness, cosmic winter, acid rain, destruction of the ozone layer, massive volcanic activity, and greenhouse warming—roughly in that order but extending across millennia. Notwithstanding the continuing dispute about whether or not this so-called *Chicxulub* impact caused a mass extinction, or whether the contemporaneous great floods of lava in India's Deccan were responsible, it seems to me that such a compendium of catastrophic phenomena appears to be perfectly sufficient to extinguish perhaps 60 per cent or more of species living at the time.

One thing that impact events, volcanic super-eruptions, and flood basalt outpourings have in common is that their deadly and far-reaching effects are a consequence of their bringing about abrupt and drastic changes in the world's climate. In this regard, there is a striking similarity with the current period of accelerating climate change, marked by an unprecedented rise in global temperatures, although in this the human race is the instigator. While operating on a far slower timescale than colliding asteroids or exploding super-volcanoes, global warming has begun to pick up pace in recent years and without preventive action will be hurtling along with the speed of an express train by the end of this century. There is no question that the current period of warming meets all the criteria needed to qualify as the fifth gee-gee; and not one that might appear on the horizon centuries or millennia down the line, but one that is real and happening now. It really is quite extraordinary that despite overwhelming evidence, vested interest, political ideology, and plain, downright ignorance still conspire to obfuscate the global warming issue—the greatest threat currently facing our planet and our race. In comparison with the tiny, ephemeral pimple that is the so-called war on terror, climate change is a great, festering sore, and one that shows no sign of healing for a very long time to come. In some ways, though, the two are comparable, and as Sir John Houghton, former Chief Executive of the UK Meteorological Office, says of climate change: 'like terrorism, this weapon knows no boundaries. It can strike anywhere, in any form—a heat wave in one place, a drought, a flood or a storm surge in another.'

As I hope to convince you in Chapter 4, there is an enormous amount we can do manage, mitigate, and ultimately solve the global warming problem, but for this to happen we must—first and foremost—recognize that it *is* a problem. Battling to get the message across is the Intergovernmental Panel on Climate Change (IPCC). Its 2001 Third Assessment Report—four weighty tomes totalling more than 3,000 pages—presents the evidence for warming from over 1,200 of the world's top climate scientists. Fighting alongside are scientific heavyweights including UK Government Chief Scientist, David King, and the aforementioned ex-head of the Met Office, John Houghton. And in the opposing corner, deriding and pooh-poohing for whatever reason, the whole idea of anthropogenic climate change and its enormous impact, is an ad hoc bunch containing a number of sincere scientists, but to a large extent hijacked by right-wing politicians, unenlightened business representatives, and self-appointed commentators, most of whom speak not from the high ground of sound technical knowledge and wisdom, but from the pit of scientific illiteracy, ideological and political dogma, and naked self-interest.

Those in the know are increasingly scared of what the future will bring. For many people, however, the obfuscation conjured up by uninformed journalists, in particular in certain tabloid newspapers, has worked, and worries over the threat of global warming barely intrude upon seemingly far more important issues, such as the goings-on in the latest television soap, or the highs and lows of a favoured soccer team. As commentator George Monbiot so perceptively observed in a recent piece on climate change in the *Guardian* newspaper:

> We live in a dream world. With a small, rational part of the brain, we recognize that our existence is governed by material realities, and that, as those realities change, so will our lives. But underlying this awareness is the deep semi-consciousness that absorbs the moment in which we live, then generalizes it, projecting our future lives as repeated instances of the present. This, not the superficial world of our reason, is our true reality.

But the future is *not* going to be same as the present—far from it. Over the last couple of centuries, industrialization and other human activities have conspired to raise the concentration of carbon dioxide in the atmosphere by over a third. Carbon dioxide is the main GHG, forming an insulating blanket that keeps the Sun's heat in and temperatures on the rise. This *greenhouse effect* is exactly the process that maintains temperatures beneath the cloud cover of our sister planet, Venus, at a snug 483 degrees C (900 degrees F). There is no prospect, at least yet, of current warming turning our own green and pleasant planet into such a furnace, but things are certainly going to warm up dramatically in the next hundred years and beyond. One of the difficulties of convincing people of the reality of global warming has arisen because until very recently it has been tricky to unravel the warming signal due to human activities from that arising from natural variations in the climate caused, for example, by changes in the output of the Sun. Total warming during the twentieth century amounted to just 0.6 degrees C: highly significant, but not enough to convince those fighting on the side of natural variation or those antagonistic to the whole idea of humans having the capability to affect nature at such a fundamental level. Only over the last decade or so have global temperatures really begun to accelerate, with nine of the ten hottest years on record all occurring since 1995, and the top four since 1997. Recently published research has shown that the last decade was hotter than any of the warm periods during medieval times, and the hottest for at least 2,000 years.

Although it is not possible to attribute a single climatic event to global warming, the broad tendency is, nonetheless, remorselessly upwards. The summer of 2003 may well, at last, have brought this trend to the attention of many, with an unprecedented heat wave across Europe seeing temperatures up to 5 degrees C higher than average across parts of the continent, and the thermometer passing 37.8 degrees C (100 degrees F) for the first time in the UK. Similar heat waves sparked the worst forest fires in Canada for 50 years, saw the highest land temperatures on record in the US, and took more than 1,500 lives in India, where temperatures reached 49 degrees C. The European summer was in fact the hottest since AD 1500, and, according to a team from the Swiss Federal Institute of Technology in Zurich, an event that would normally be expected only every 46,000 years. What is perhaps most extraordinary about the baking European summer of 2003 is not the unparalleled wildfires that raged across Portugal, Spain, and France, nor the 15 billion US dollars' cost to agriculture, but the enormous death toll. In all, over 35,000 lives are estimated to have been lost to the sweltering conditions, including over 15,000 in France, 7,000 in Germany, and 2,045 in the UK—in total, more than ten times the number killed in the World Trade Center terrorist attack and, in the UK, two-thirds of the lives lost every year on the roads. Because the deaths were spread across a continent and because many people died alone, far from the view of the cameras, nobody wants to know. As a consequence of the heat wave catastrophe no war has been declared on global warming; neither has the US—the world's greatest polluter by a mile—curbed its arrogance and indifference and embraced the need to solve the climate change problem.

In fact, it is business as usual on planet Earth. The Kyoto Protocol, proposed in 1997 to reduce GHG emissions to 5.2 per cent below 1990 levels by 2008–2012 only came into force in February 2005. Hindered by the refusal of the US, Australia, and Canada to ratify the agreement, only the surprise announcement in September 2004 by Vladimir Putin that Russia was to sign up saved the treaty from oblivion. The situation remains far from rosy. Despite early and enthusiastic adoption and ratification by the European Union, which overall looks as though it is meeting the fall in emissions required by the protocol, even here there are problems, and big rises by a few states threaten to counteract the best efforts of those, such as the UK and Germany, that are closer to meeting their targets. Spain's emissions, for example are over 33 per cent up on 1990, with Austria, Denmark, Ireland, Italy, and Portugal also struggling to meet their targets. Among those ignoring the protocol the situation is just as bad, with Australia seeing an 8 per cent rise, the US a 13 per cent increase, and Canada a 15 per cent hike in GHG emissions since 1990. The US is without doubt leading the way towards our hothouse future, with just 4 per cent of the world's population pumping out more than a fifth of all greenhouse gases. If the US states were independent nations, 25 of them would be in the top 60 highest emitters of greenhouse gases, with Texas alone surpassing France. Taken overall, emissions of carbon dioxide have actually gone up by 10 per cent since the Kyoto meeting in 1997. In 1663, before the Industrial Revolution began to build a head of steam, the carbon dioxide concentration in the atmosphere stood at 279 ppm. By 1970 this had risen to 325 and by 2004 to 378. And the rate of increase itself is on the rise, from an average of around 1.5 ppm over the last few decades to 2.5 ppm since 2001. This acceleration probably reflects the wholesale burning of coal to provide the electricity needed to power China's rapidly expanding economy, a policy that the World Energy Council predicts will raise carbon dioxide emissions by a third in a little over 15 years.

Assuming, then, that we continue to do nothing, what can

we expect to be the consequences of climate change over the next hundred years and beyond? The fundamental predictions are summarized in the IPCC Third Assessment Report. Global average surface temperatures are expected to rise by between 1.4 and 5.8 degrees C, with the most likely rise being around 3 degrees C. At the same time, melting glaciers and ice caps and the thermal expansion of the oceans are forecast to lead to a sea-level rise of 9 to 88 cm, with around 40 cm seen as most plausible. These figures, however, provide only a glimpse of the true picture and say nothing about the ramifications of such variations in temperature and sea level for our society and the environment. Furthermore, they mask regional and local effects that may be far more extreme. Things have also moved on since publication of the IPCC report in 2001, with new research pinning down predictions more accurately and drawing attention to previously unforeseen consequences of warming, and observations highlighting the accelerating rate of climate change today. Scariest of all are the findings of a workshop of top atmospheric scientists held in Dahlem, Berlin, in June 2003. The meeting addressed the so-called *parasol effect*, the veil of soot, smoke, and other particles in the atmosphere that counteracts warming by reflecting incoming solar radiation back into space. The workshop concluded that over the last century this detritus of industry had actually shielded us from three-quarters of the effects of global warming. In other words, given a sparklingly clean atmosphere, global temperatures would have risen by almost 2.5 degrees C since 1900 rather than the observed 0.6 degrees. As fossil fuels are used up and industries become cleaner during the course of the next hundred years, so too the atmosphere is expected to return to a more pristine condition. Tremendous news for the occupants of the great industrial cities, but not for the human race as a whole. Without a reflective layer of pollutants, temperatures are predicted to soar far above the levels forecast by the IPCC report, with worst-case scenarios visualizing a rise of 7–10 degrees C by 2100. Such a huge change over such a short period of time would be devastating for the Earth and all life upon it, leading to mass extinctions of animal and plant species, desperate problems in food production and water supply, the collapse of many economies, and drastic changes in every aspect of our lives.

But enough of the future; we can already see the impact of global warming all around us, if we choose to look. The past hundred years has seen an unprecedented retreat of glaciers all over the world, and in the Alps around half of total glacier mass was lost between 1850 and 1990. In the US, the 150 glaciers that gave a name to Glacier National Park have been culled to just 50, while the ice caps of Patagonia are melting so fast that they have generated a deluge of up to a thousand cubic kilometres of freezing water over the last half-century. In Africa, most glaciers will have disappeared in another two decades, including those in Uganda and the Congo that feed the headwaters of the Nile. Furthermore, glaciers on Kenya's Mount Kilimanjaro have shrunk by more than 80 per cent since 1912, and this iconic landmark is predicted to be ice-free in just over ten years' time. Inexorably rising temperatures are already beginning to pose a threat to the inhabitants of mountainous regions, and in 2003 ice was melting at such a rate on Mont Blanc that the Alps' highest peak had to be closed to climbers. In the Himalayas, rapidly retreating glaciers have contributed to the formation of 40 huge lakes in Nepal and Bhutan, which are expected to burst their banks in the next five years or so, sending torrents of icy water cascading into inhabited valleys below. Meanwhile, hundreds more glacier-fed lakes continue to expand and deepen, threatening towns and cities in Tibet, China, Afghanistan, Pakistan, and northern India.

In many upland areas, it is only the permanently frozen rock and soil that is keeping mountains from falling apart, but this is also now beginning to melt. In the Alps, temperatures have risen by 2.1 degrees C since the 1970s, melting the binding permafrost at higher altitudes and contributing to greater numbers of rock falls, landslides, and mud flows. Things are likely to get considerably worse, however, with the potential increasing for the failure of entire mountainsides and the resulting burial of alpine villages and holiday resorts under billions of tonnes of rock. Certainly the changing climate is already making those with a vested interest in winter sports quake in their ski boots. Two of the four Scottish ski resorts are already up for sale, as a lack of snow makes them unviable, while a UNEP (United Nations Environment Programme) report published in December 2003 forecasts that the snowline in Austria will rise by up to 300 m in the next 30–50 years. Over the same period it predicts that the number of Swiss resorts with reliable snow will fall by a quarter, with up to half of all the country's winter sports resorts facing bankruptcy or economic hardship.

At the poles, signs of accelerating warming are even more apparent, and both the Arctic and Antarctic regions are heating up more rapidly than the rest of the planet. On the Antarctic peninsula, green plants are now far more common than when Scott and Amundsen battled for the pole, and with temperatures up by 2.5 degrees C since 1940, the gigantic floating ice shelves that border the peninsula are in full retreat. Between the middle of the twentieth century and 1997, around 7,000 square kilometres of ice shelf—an area more than twice the size of Luxembourg—became detached, broke up, and floated into the South Atlantic, forming armadas of giant bergs. In just the last five years, however, another Luxembourg-sized chunk has broken away, nearly all of it in the form of the Larsen B Ice Shelf, which took just 31 days in 2002 to separate itself from the Antarctic continent and fragment into thousands of icebergs. The Larsen Ice Shelf proper—a mass of ice 300 m thick and about the size of Scotland—is also melting at an unprecedented rate and pouring an extra 21 billion tonnes of freezing water into the southern ocean every year, not far off eight times the annual flow of the River Thames. With ice shelves covering 50 per cent of the Antarctic coastline, there is enormous concern that wholesale melting will lead to their break-up, exposing the great ice sheets grounded on the seabed and on the continent itself. There are fears that the West Antarctic Ice Sheet, in particular, might be vulnerable; its collapse and fragmentation having the potential to cause a

5–6 m rise in sea level, sufficient to drown all the major coastal cities.

The Arctic too is suffering, with sea ice thinning by 40 per cent over the last 35 years and its area shrinking by 6 per cent since 1978. In the first year of the new millennium the North Pole was ice-free in summer, a feature that is likely to become increasingly common in future decades as global warming really hits home. Seymour Laxon, a climate physicist at University College London, showed recently that just a single day's extension to the summer melting season can cause the ice to thin by 5 cm, a particularly worrying scenario, given that the melting season is lengthening by an average of 5 days every decade. Even the great ice sheet that covers most of Greenland in a frozen shroud up to 3 km thick is no longer safe, with the first tentative signs of melting being picked up. Further new research published in the journal *Nature* by Jonathan Gregory, of Reading University, and colleagues provides a bleak outlook for the ice sheet, which appears doomed to melt in its entirety—probably never to re-form— unless there are drastic cuts in GHG emissions.

It should not be surprising that the most profound effects of global warming are manifest where temperatures are at their lowest, but the impact of climate change is now seen and felt far and wide, and in ways that are sometimes unexpected and unforeseen. In just the last 30 years the northern hemisphere growing season has lengthened by 11 days, and in temperate zones, such as the UK, broadleaf trees may start to keep their leaves over winter. Aesthetically pleasing perhaps, but raising the risk of their being toppled by storms, which

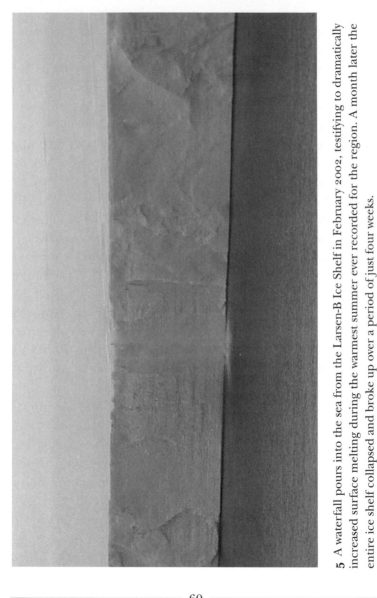

5 A waterfall pours into the sea from the Larsen-B Ice Shelf in February 2002, testifying to dramatically increased surface melting during the warmest summer ever recorded for the region. A month later the entire ice shelf collapsed and broke up over a period of just four weeks.

are also predicted to become more common and more powerful. In fact, according to the UK Met Office, Britain is already twice as stormy as it was 50 years ago, although a longer period of observation will be needed to confirm if this trend continues into the future. While it is simply not possible to pin the blame for any single weather-related natural disaster on global warming, the trend towards bigger and badder weather events is becoming increasingly apparent. Incidences of extreme precipitation rose by up to 4 per cent at mid to high latitudes during the second half of the last century, and more windstorms, floods, and rainstorms are forecast. The jury remains out on whether or not a warmer world will see more hurricanes, typhoons, and tornadoes, but a number of studies predict that such storms will become more powerful and increasingly destructive.

Where wind leads waves are rarely far behind, and evidence is already accumulating for bigger and more powerful waves. Around the southern and western coasts of the UK, average wave heights are now a metre higher than they were in the 1970s, while the largest waves have increased in height by a third to an alarming 10 metres. A combination of more powerful storms, higher waves, and remorselessly rising sea levels can be expected to raise dramatically the threat of coastal flooding and even the drowning of entire nations. Notwithstanding melting of the Greenland or West Antarctic Ice Sheets, the future looks bleak for the Maldives in the Indian Ocean, which a mere 1 m rise would see inundated. With its 9,300 inhabitants crammed onto just 75 acres of land barely a metre above sea level, the South Pacific island of Tuvalu is certain to suffer the same fate. Already 3,000 former inhabitants live abroad and 75 are relocated every year, but the government is now looking seriously at the feasibility of evacuating everyone from the island before the situation becomes critical. Elsewhere, all countries with a coastline are at risk, with 200 million people likely to be affected by coastal flooding by 2080. A combination of rising sea levels and subsidence is predicted to result in the loss of 16 per cent of the land area of Bangladesh, which supports 13 per cent of the population. In Egypt, sea-level rise threatens the heart of the economy, with more than half of the country's industrial capacity located within 1 m of the current sea level. Prospects are no better in the developed world, with rising sea levels threatening 200 billion pounds' worth of assets and 2 million homes in the UK, and the US predicted, by the end of the century, to lose an area of land to the sea equal to the states of Massachusetts and Delaware combined.

Water—either too much of it or too little—is, in fact, fast becoming recognized as one of global warming's most disruptive bequests. Most climate models forecast that as the world hots up, current climatic characteristics will be enhanced, so regions that are now dry will suffer from prolonged drought, while those that are already wet will simply get wetter. Australia, therefore, will face increasingly frequent baking summers, while the deserts of North Africa will creep inexorably towards southern Europe, where blazing summers will make it far too hot for a beach holiday. In contrast, northern Europe and the UK will experience mild and wetter winters and the prospect of flooding becoming a perennial issue. By 2080, the number of people at high risk of flooding in the UK may well have risen to 3.6 million, with floods that now return every century or so making an appearance every few years. In many regions, water will become such a scarce resource that it will become a trigger for war. Already 2.4 billion people lack adequate sanitation and over a billion have no water in their homes. By 2025, over half a billion people will have no access to clean water, and the UN identifies 158 international river basins that could become future flash points as governments struggle to provide water for their citizens, industry, and agriculture.

Today's terror-riven world might appear scary and insecure, but things will be far worse as climate change starts to take a tighter hold. The CIA—hardly renowned for sticking its neck out—anticipates a future racked by social and economic instability and civil strife, with wars and mass migration disrupting the fabric of nations in many parts of the world. By the middle of this century, global warming is likely to have displaced around 150 million people, with cross-border migration of millions of refugees triggering social volatility, feeding racism, resentment, and bitterness, and providing fertile ground for terrorist recruitment. Disease will be rife among both displaced and indigenous peoples, with a million extra cases of malaria a year already attributed to climate change, the higher soil moisture after floods allowing insects to survive for longer. In coming decades, the malaria threat is expected to move northwards into temperate regions, and into upland areas that were previously too cool. Diarrhoea and malnutrition will follow hard on the heels of malaria, with the World Health Organization announcing that global warming can already be held directly responsible for 150,000 deaths a year due to these three biggest killers in the developing world. Even industrial countries are far from immune, with reported cases of salmonella food poisoning in the UK increasing by 12 per cent for every 1 degree C rise in temperature.

If climate change affected only the human race, then it might be reasonable to take the view that any consequences we suffer are little more than we deserve. Unfortunately, no single animal or plant on the planet is impervious to the unprecedented warming that is currently taking place. The Earth has seen dramatic swings in temperature throughout its history, and no more so than during the Ice Ages, the last of which ended barely 10,000 years ago. Never before, however, has climate change been as rapid and dramatic as we see today, with the result that there is simply no time for many plant and animal species to adapt. As a consequence, it is appearing ever more likely that we are living through the sixth great extinction in the Earth's 4.6 billion-year history. The previous five arose from natural events, such as the asteroid impact and mass outpourings of lava that wiped out the dinosaurs and 60 per cent of all species. Blame for the current mass culling, however, sits foursquare on the shoulders of the human race and is occurring at between 1,000 and 10,000 times the natural rate. The numbers involved are truly staggering, with Professor Chris Thomas, a conservation biologist at Leeds University, and colleagues forecasting that human activities—through climate change—will drive a quarter of land animals and plants into extinction by 2050, over a million species in all. By 2100 the Earth may be warmer than for 10 million years, with conditions prevailing comparable to those that existed before the bulk of today's species and plant and animal communities had evolved or developed.

I shall stop there, before the urge to head for the bathroom and open my veins becomes too strong. If like me, this rummage through the gee-gee portfolio has given you a fit of the heebie-jeebies, then take heart from the fact that succeeding chapters will, I very much hope, lighten the mood. We have always been at risk from great volcanic blasts, giant waves, or careering rocks from space, so the situation in these respects is currently no better and no worse than it was a hundred or a thousand years ago. In fact, it can justifiably be argued that our exposure to such phenomena is actually reduced as we now possess the capabilities to forecast such events, and even, in some circumstances, to prevent them. Admittedly, accelerating climate change is an enormous and growing problem and one that is not going to go away. It would be foolish to suggest that it can be stopped in its tracks (more about this in Chapter 4), but there are ways to manage and mitigate its impact, to adapt to the changes it will undoubtedly bring, and—ultimately—to bring it under control. Things on planet Earth at the start of the third millennium are bad, but not so desperate that we have to give up all hope just yet.

2

Defending the Earth

'Gracious goodness me!' said Chicken-licken, 'the sky must have
fallen; I must go and tell the King.'

Young Folks' Treasury (1919).

The evening air at Kitt Peak, high above the Arizona
desert, is rarefied, so much so in fact that you can
quite easily imagine yourself at the edge of space,
where oxygen is a luxury and every breath a struggle. And
space is very much the focus of Kitt Peak, for here a gleaming
cluster of 18 silver and white domes houses a range of major
optical telescopes for the study of the Sun and the night sky.
On a chilly evening in June 2001, I have come to this spectacular mountain-top to try a bit of asteroid spotting, because
one of the Kitt Peak National Observatory's telescopes forms
the business end of the University of Arizona's Spacewatch
Project, dedicated to finding Earth-threatening objects
before they find us. The technology on show here is a curious
mix of twenty-first-century state of the art and early twentieth-
century decrepitude. The Steward Observatory dome houses
a 0.9-m telescope that has certainly been around a bit.
Installed in 1921 on the University's Tucson campus, 70 km
away, this workhorse was moved to Kitt Peak in 1962 to take

advantage of the better conditions for observation. Two decades later, having been used to discover the first optical pulsar in 1969, the venerable telescope was flagging and fell into disuse until it was resurrected by Spacewatch in 1982. Then the newly formed organization managed to persuade Kitt Peak's director to let its members have exclusive access to the telescope, purely on the understanding that they guaranteed to refurbish and maintain the sexagenarian optics and machinery.

On a summer evening 20 years later, the telescope remains ropy enough to require kid-glove treatment, and it is now far too doddery to be allowed to track smoothly across the night sky. Instead, the instrument is simply pointed through the rectangular opening in the rather creaky dome and left to scan the exposed strip of sky as the Earth itself rotates beneath it. The telescope's exterior may be a cast iron and brass hangover from a bygone age, but its innards are far more sophisticated. No longer does an observer lounge in a tilted chair with his eye glued to the instrument's eyepiece. In 1983 Spacewatch installed an electronic imaging detector system based upon a charged-couple device or CCD, a piece of gadgetry that is able to build up a detailed image—even in very low light conditions—by converting incoming photons to electrons that can be digitally processed. The same technology is used in TV cameras and is now filtering into top-of-the-range camcorders.

As darkness fell, observer and research specialist Arianna Gleason took me down to the telescope control room to start the night's search. After setting the exposure and positioning

6 The 0.9 m Spacewatch Telescope in the Steward dome (left) at Kitt Peak National Observatory, Arizona, is dedicated to spotting potentially threatening Near Earth Asteroids (NEAs). As of January 2005, 3,167 NEAs had been spotted and their orbits determined, including 757 monsters more than 1 km across.

the instrument for the first scan we sat down in front of two large computer screens which—as the CCD built a picture— soon lit up with spectacular views of the universe: not just the pin-prick stars that adorn the our city skies, but the tiny spirals of distant galaxies, the smoky clouds of interstellar gas, and even a streak or two of light as a satellite sped across the field of view. There was so much to see that it was difficult to imagine that I was staring at a minuscule area of the night sky, just half a degree across and seven degrees long. On a good night the telescope was capable of picking up a good 300–400 asteroids in a single 90-minute scan, so where were they? They were there somewhere, but their movement against the backdrop of distant stars was so slow that it would not become apparent until the same region of sky had been scanned for a second time and the two series of images compared. The ancient telescope creaked and groaned as it prepared to scan the tiny sliver of sky for a second time. The first pass had taken a tedious 30 minutes, but now the fun was about to start. As images of the same patch of sky started to appear again on the monitors, small coloured boxes and circles started to materialize around innocuous-looking points of light, accompanied by mystifying alphanumeric designations. Using extraordinarily complex software that took over eight man-years to develop, the computer was automatically picking out objects that had moved in the intervening period between the two scans, and using a catalogue of the orbital characteristics of known objects in its memory to suggest what they might be. Amazing as this all was, I was rapidly losing interest as the computer seemed to be having all the fun, and my fingers wanted to find out just how good this motion-detection routine was. Might it be missing something? Using a mouse, I clicked randomly on the screen, bringing up a more detailed image of this patch of sky, and flipped between it and an image of the identical region of space captured during the first scan. Immediately, something jumped out at me. One 'star' had clearly moved between the two scans. As I excitedly drew Ariana's attention to this astonishing—or so it seemed to me—discovery, she seemed distinctly unimpressed, informing me in a rather blasé manner 'Yup—that looks like a new asteroid.' I was stunned. Could it really be that easy? Well, perhaps not.

First of all, details of the new object—provisionally designated B6B11V—would be sent to the International Astronomical Union's Minor Planets Center in Massachusetts, which would then confirm whether or not it was really new, attempt to determine its orbit, and assess whether or not there was any risk of it threatening the Earth. But surely, I thought, this was a real find for Spacewatch—a new asteroid. Not really, I quickly found out, for since it started its sky survey in 1984, the Spacewatch telescope had achieved a total of 350,000 asteroid detections (a figure that had reached more than 800,000 by the end of 2003), around 45,000 of which were new. One more was no big deal. Ultimately, it proved that the deal was even smaller than first thought, with the Minor Planets Center reporting that McGuire's asteroid was not even new. In fact, B6B11V was actually 2000E7167—an asteroid originally spotted the previous year but lost again before its orbital characteristics could be pinned down. But could it be the 'big one', the asteroid with the Earth's name on it? With a diameter of 5 km it was certainly massive enough to make a very large hole in our planet and, at the very least, catapult our society back to the Dark Ages, but with the object approaching no closer than 400 million km, the chances of this ever happening were effectively zero. With all possibility of fame—or would that be notoriety—extinguished, I would have to console myself with the words of Brian Marsden, Director of the Minor Planets Center, who has described McGuire's asteroid as 'somewhat interesting'.

Sky surveys like Spacewatch make up our first line of defence in the battle to stop objects from space colliding with our planet. Together with other systematic searches such as MIT's LINEAR (Lincoln Near-Earth Asteroid Research), the NASA Jet Propulsion Laboratory's NEAT (Near-Earth Asteroid Tracking), and the Flagstaff (Arizona) Lowell Observatory's LONEOS (Lowell Observatory Near-Earth Object Search), Spacewatch has spotted more than 650 *Potentially Hazardous Asteroids*—objects bigger than around 150 m that have the potential to make threateningly close approaches to the Earth. Since the spectacular collisions between the disrupted Shoemaker-Levy comet and the planet Jupiter focused minds so effectively in 1994, the discovery rate of *Near Earth Asteroids* (NEAs) has accelerated enormously. In 1995 only 333 NEAs had been identified, 192 of which had diameters of a kilometre or more. By early 2005 these numbers had risen respectively to 3,167 and 757. Things have come a long way since David Morrison of NASA's Ames

Research Center in California hazarded the guess that there were probably fewer people involved in hunting for Earth-threatening asteroids than worked in a McDonald's restaurant. The occupation has also accrued an air of respectability within the scientific community and is no longer regarded as the astronomical equivalent of train-spotting. This in turn has helped to bring about increased levels of funding, particularly through NASA, leading to better detection equipment and a concomitant rise in catalogued NEAs.

A number of new surveys are on the cards to increase detection rates even further, including Pan-STARRS, which will begin operations at the University of Hawaii's Institute for Astronomy in 2007. The Panoramic Survey Telescope and Rapid Response System combines four telescopes—each with a three-degree field of view—to provide the capacity to scan the entire sky (visible from Hawaii) several times every month. Pan-STARRS will be dedicated primarily to the search for threatening asteroids and comets and is likely to push up significantly the number of catalogued objects. Following on the heels of Pan-STARRS is LSST, the proposed Large-Aperture Synoptic Survey Telescope, sometimes referred to as the 'dark matter telescope'. This huge telescope—8.4 m across—is planned to be 20 times more powerful than any other astronomical observatory already operating or under construction. Not only will it be capable of plumbing the depths of the universe looking for the mysterious and elusive dark matter, but its unmatched optics will also make the LSST an invaluable addition to the stable of asteroid- and comet-spotting telescopes. Should it be built, the LSST will have the capability to spot every Near Earth Object down to 300 m in size, a tremendous feat if it can be accomplished.

Certainly the most entertaining advance in the search for potentially hazardous asteroids and comets involves the Spacewatch project's Fast Moving Object (FMO) public program. This utilizes volunteers with PCs and web access to analyse images taken by one of the project's Kitt Peak telescopes to search for asteroids moving so close to the Earth that their passage is marked by streaks of light. It turns out that many of these streaks are simply too faint for software to distinguish them from background noise, and careful inspection with the human eye proves to be by far the best method for distinguishing these cryptic trails. As this is a very time-intensive business, Spacewatch has begun a recruitment drive to find amateur enthusiasts who are keen to help. No professional experience is necessary, and while some knowledge of astronomy would be useful, the main qualifications are an interest in asteroid hunting and a regular availability to review image data. So if you fancy doing your bit to protect the Earth, check out the Spacewatch FMO site at http://fmo.lpl.arizona.edu/FMO_home/index.cfm.

The hunt for potentially threatening asteroids forms perhaps the most important element of the multinational *Spaceguard* initiative, launched in Rome in 1996 with the stated aim of protecting the Earth environment against bombardment from comets and asteroids. In 1998, NASA's contribution to Spaceguard began with a commitment to ensure that 90 per cent of all asteroids with diameters of a kilometre or more were identified and tracked by the end of 2008. Through supporting the main sky surveys like Spacewatch, LINEAR, and NEAT, about 60 per cent of this task has now been accomplished, assuming that the total number of NEAs of this size is between 1,000 and 1,200, which is the current consensus. The risk of a large asteroid striking the planet by surprise has thus already been dramatically reduced, although it may take many decades before the final 10 per cent of NEAs are discovered, and even then we can never be absolutely certain that all have been accounted for. Furthermore, there is a strong possibility that the number of potentially threatening asteroids orbiting the Sun in the vicinity of the Earth may fluctuate over time, due to disturbances in the heavens.

Although the structure of the solar system may seem inviolate, this is far from the case, and the status quo has been frequently disrupted in the past and will be again. Most impact scientists hold the *uniformitarian* view that the rate of asteroid and comet impacts is constant over time; some, however, disagree. British astronomers Victor Clube and Bill Napier, who head the self-styled 'coherent catastrophist' school, argue for impact clustering and the existence of periods in Earth history, some quite recent, when collisions with extra-terrestrial bodies were far more common than they are today. Clube, Napier, and colleagues account for these more active periods in terms of the arrival—perhaps every 20,000 to 200,000 years—of a new, large comet in the inner solar system. For one reason or another, maybe through getting too close to the Sun or to Jupiter's strong gravitational pull, the comet is captured and—like Shoemaker-Levy—torn apart. Over thousands of years the countless fragments of rock spread out along the comet's orbit, forming a ring of debris just waiting for something to hit. A large comet, when broken up in this way, may 'seed' the inner solar system with many thousands of chunks of rock a kilometre or more across, each capable of triggering a global catastrophe should it strike the Earth. The coherent catastrophists claim that such a giant comet entered the solar system as recently as 20,000 years ago, fragmenting soon afterwards to form a trail of debris that encounters and pummels our planet every few thousand years. The last major barrage is alleged to have happened just over 4,000 years ago, causing Bronze Age mayhem and the demise of budding civilizations across the planet. The next may be along in just another millennium.

Almost as worryingly, Michael Ghil and co-workers at the University of California, Los Angeles, have recently drawn attention to the fact that our neighbouring planets may also have a periodically destabilizing effect on the rate at which extra-terrestrial objects strike the Earth. Rather than following perfectly predictable and unchanging trajectories about the Sun, Ghil and colleagues discovered that the orbits of Jupiter and Saturn—the solar system's largest planets—were actually poised on a knife-edge, with the slightest variation having the potential to send them into chaos. Ensuing wild variations in their orbits would have the potential to trigger instability in the main asteroid belt between Mars and Jupiter, where a flotilla comprising up to 2 million chunks of rock provides an enormous source of spare ammunition for a future bombardment of our planet. Disturbed by changes in

the orbital patterns of the two giant planets, main belt asteroids could easily be dislodged from their orbits and sent
Earthwards. While still new, the work of Ghil and his co-
workers provides some compelling evidence for these sudden
bursts of madness in the solar system, with their sophisticated
computer model predicting episodes of chaos 65 and 250
million years ago, both times when impact events have
coincided with mass extinctions on Earth. The next period of
interplanetary havoc is predicted to occur in around 30 million years' time, so our descendants have quite a while to
work out a way of fending off this particular fusillade. The
lesson that the work of Ghil's team and the coherent catastrophists holds for us, however, is that numbers of NEAs are
likely to vary over time. We cannot sit back and relax once we
think we have spotted all those that constitute a threat, for
the population may be refreshed. In other words, 'keep
watching the skies' is a slogan that will be as relevant to our
distant descendants as it is to us today.

Watching and cataloguing are all well and good, but
already there is a growing thirst to get out there and discover
more about what makes asteroids and comets tick, and in
particular, what they are made of and how they are put
together. After all, if we are ever going to have to give one a
nudge, we need to know what method to use and how the
object is likely to respond. The one thing we don't want to
accomplish is the shattering of a large chunk of rock into
several smaller but still dangerously big lumps, without changing its trajectory. This would simply result in the Earth
being clobbered by something akin to a shotgun blast rather

than a single rifle bullet. A number of unmanned spacecraft have already reconnoitred asteroids and comets from close up, while a whole fleet of more sophisticated US, European, and Japanese probes are either already *en route* or at least off the drawing board. The most spectacular mission so far involved the launch of the NEAR (Near-Earth Asteroid Rendezvous) probe in 1996. Barrelling past asteroid 243 Mathilde 18 months later, and asteroid Eros in 1998, the spacecraft returned to circle Eros in 2000, eventually accomplishing a spectacular and impromptu soft landing on the surface in early 2001. Renamed *NEAR Shoemaker* during its journey, to commemorate the death in 1997 of planetary geologist Eugene Shoemaker, the probe's cameras returned a series of stunning high-resolution images of the asteroid's surface, revealing a number of very large craters formed during collisions with smaller objects. Most importantly, the existence of these craters suggests that Eros is really just a pile of agglomerated rocky rubble rather than a seamless chunk of solid rock, the thinking being that a rubbly asteroid will be able to absorb more energy before shattering than its monolithic equivalent. Imagine bashing a pile of gravel with a hammer rather than a single brick. If other asteroids are constructed in the same way, then such information will prove invaluable when and if we have to resort to force to ward off an asteroid heading in our direction. Because asteroids and comets differ so widely in their make-up, however, more than one mission is needed to pin down their characteristics. As NASA put it on their Near Earth Object Program website (http://neo.jpl.nasa.gov/index.html), if we need to nudge an asteroid out of an earth-threatening trajectory, 'it makes a big difference whether we're dealing with a 50 m sized fluff ball or a one-mile slab of iron.'

Comets, with their often unexpected *apparitions* through history and long, gleaming tails pasted across the sky, have always seemed very different bests from asteroids, somehow altogether more flamboyant and exciting. But at heart, just how different are the two. In January 2004, in an attempt to find out, the NASA *Stardust* probe hurled itself through Comet Wild's *coma*—the glowing envelope of gas and dust enclosing the rocky nucleus. The probe's cameras revealed a body pockmarked with craters that looked remarkably like an asteroid. There, however, the resemblance ends, with the comet displaying activity far beyond anything demonstrated by an asteroid. Jets of gas, formed as the Sun's heat vaporized ice within the nucleus, blasted out chunks of rock as big as boulders and sent trails of dust and gas millions of kilometres long into space. While *Stardust* miraculously dodged the larger chunks of debris, it took a battering from smaller dust and sand-sized particles, which breached the first layer of shielding in several places but failed to cause any serious damage. During its roller-coaster ride, *Stardust* accomplished its main objective of capturing cometary dust samples, which it will return to Earth in 2006. Not to be outdone, the European Space Agency launched its comet-chasing *Rosetta* probe into deep space in March 2004, charged with going into orbit about Comet Churyumov-Gerasimenko before sending a lander down to the surface. We will have a bit of a wait before the action starts, however, as *Rosetta* first has to swing round the Earth three times and Mars once, and scoot past asteroids Steins and Lutetia before catching up with the comet in 2014. Two other missions worth a mention are *Dawn*, which sees a US spacecraft heading off in 2006 to orbit two of the largest asteroids, Ceres and Vesta, and *Muses-C*, a Japanese probe already on its way to asteroid 25143 Itokawa to take samples and bring them home in 2007.

By far the most exciting mission in the offing—at least from the perspective of countering the impact threat—is NASA's *Deep Impact*, named after the Hollywood blockbuster that catalogued the effects of a comet fragment striking the Atlantic Ocean. Launched in January 2005, *Deep Impact*'s main task centres around the ultimate exercise in target practice: the launch of a high-speed projectile at a comet, designed to find out how the chunk of rock and ice will respond when struck. As the probe hurtles past Comet Tempel 1 on July 4th, 2005, a *smart impactor* will detach itself and head off on a trajectory designed to intersect that of the comet. Twenty-four hours later, the 360 kg copper missile will close on the comet, taking pictures all the time, before crashing into its sunlit side at a relative velocity of 10 km a second—over 15 times the speed of Concorde. The collision will be filmed by the main probe and is expected to result in a fresh crater bigger than a football field and as deep as a seven-storey building. Observation of the cratering process and its legacy will, it is hoped, provide vital information on the strength and structure of the comet, which may be utilized should plans ever have to be drawn up to attack a threatening comet in anger.

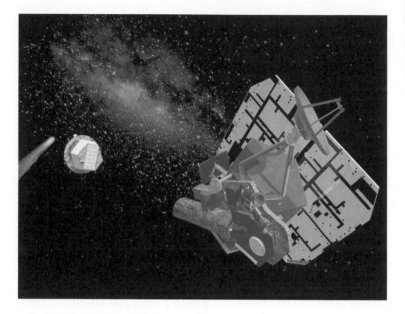

7 Launched in early January 2005, the *Deep Impact* probe is expected to reach the vicinity of Comet Tempel 1 on July 4th, 2005. In order to examine the response of the comet and to learn more about its structure, a projectile will be sent on a collision course, striking the surface with a velocity 15 times greater than Concorde, and excavating a crater 25 m deep and more than 100 m across.

So, just how big a threat to the Earth are comets? Compared to potentially hazardous Near-Earth asteroids, many comets are colossal, comprising aggregations of rock and ice up to 100 km or more across. Rather than follow the broadly circular and largely predictable orbits of the NEAs, comets follow strongly elliptical paths that can take them well beyond the edge of the solar system and deep into interstellar space. Far from the Sun, they are difficult to spot, but as they encroach on the inner solar system they undergo a seemingly miraculous transformation as sunlight boils off gas, dust, and debris from the central nucleus, forming extravagant 'tails' that may extend across space for 100 million km or more. As a comet hurtles Sunward from deep space, so it achieves velocities in the range of 60–70 km a second—three times the speed of the typical NEA, and close to a hundred times faster than the Anglo-French aerial masterpiece. This is bad news for any other body—such as the Earth—that gets in the way, because a collision would be far more energetic and therefore much more destructive than any caused by an asteroid. There is some good news, which is that comets are far less common in the solar system than asteroids, and collisions with them are therefore far less frequent. On the other hand, the appearance of a comet with our name on it may not be predictable. While Halley's Comet, the most famous of all, reappears like clockwork every 76 years, and has been observed on dozens of occasions over thousand of years, others may not be so well behaved. Extrapolating ahead, it is possible to determine that—at least until AD 3000—Halley will not even come close to threatening our planet. This exercise cannot, however, be applied to the so-called *parabolic* comets, those that crawl along gigantic orbits that take them on immeasurably long journeys far beyond the fringes of our solar system. Some parabolic comets may return to the solar system every few thousand years and may have been observed, if not recorded, by our distant ancestors. Others may never have ventured into the inner system before. With no earlier visitations to help predict an orbit, our first view of one of these objects heading our way may leave us just six months to prepare for an unavoidable and calamitous colli-sion. Furthermore, a comet on its first tour of the solar system is likely to be the father (or mother) of all comets: a 100-km wide assassin unaffected by previous battles with the hurricane of solar particles known as the *solar wind* that will, over time, erode a comet's mass.

We now have a fair idea of how many asteroids and comets constitute a potential threat, at least in the immediate future, and we are beginning to understand a little more about their form and structure. But let's get to the nitty-gritty of the problem. What can and should we do if we detect an extra-terrestrial body on an apparent collision course? Extraordin-arily, astronomers were faced with this very question early in 2004 when, for several hours on the night of 13 January, it looked as if there was a one in four chance of asteroid AL00667 (later redesignated 2004AS1) crashing into the northern hemisphere. Although not large enough to cause global devastation, at 30 m across the object would have had the capacity to wipe London or Washington from the face of the map. In the end no collision ensued, but as a foretaste of what we or our descendants may have to face in the future, it is worth looking in a little more detail at the circumstances of the crisis and devoting a little space to consideration of the lessons the incident has taught us.

Coincidentally, the perceived emergency started just 36 hours before George Bush's much heralded speech at NASA headquarters, announcing a new initiative for the manned exploration of the Moon and Mars. In New Mexico, a telescope attached to the LINEAR sky survey routinely recorded four images of a moving object and sent them to the Minor Planets Center in Massachusetts. Here, researcher Tim Spahr ran the observation data through a computer program to provisionally determine the future behaviour of the object and—before heading off for dinner—posted the information on the open-access Near Earth Object information web page (NEOCP). This would allow both professional and amateur astronomers to build on the LINEAR observations and continue to track the object that night. Within an hour of the posting, German amateur astronomer Reiner Stoss checked out the prediction and was stunned to find that asteroid ALoo667 was forecast to get 40 times brighter over the course of the next day. This meant that by then the object would be six times closer to the Earth; in essence, the asteroid was heading our way—and fast. Unable to keep the news to himself, Stoss posted his concern about the 'bogie' on the Minor Planets Mailing List chat room, which just happened to be picked up by professional asteroid researcher Alan Harris at Boulder, Colorado's Space Science Institute. Calculating that the object could hit in just a day's time, Harris immediately contacted some heavyweight colleagues, including Don Yeomans, head of the NEO Program Office at NASA's Jet Propulsion Laboratory in Pasadena (California), and NASA Ames Research Center's David Morrison, Chair of the International Astronomical Union's NEO Working Group.

After an hour and a half's wait, the NASA NEO Office managed to contact Brian Marsden, Director of the Minor Planets Center, who at that time knew nothing of the apparent crisis. Marsden made some quick calculations based on the original data and came up with an alternative path that showed the object actually receding from the Earth, and this was posted on the website to replace the original prediction. Soon after, having finished his meal and logged on at home, Tim Spahr also became aware of the growing panic. He too made a recalculation of the asteroid's path, this time showing the object once again heading towards the Earth, but narrowly missing rather than colliding, and posted it on the web in place of Marsden's. The situation was a big embarrassment for the Minor Planet Center which had, it seemed, issued the first ever short-term prediction of an asteroid impact without noticing it. Of course, the circumstances were not quite this straightforward. The prediction was based upon just four observations made in a single night, and as a consequence significant uncertainties were attached to the prediction of the object's trajectory, which is how Spahr, Harris, and Marsden were able to derive different results from the same data. In the end, speculation was ended in a relatively straightforward manner. If asteroid AL00667 was really on a collision course, then it was possible to use information about its trajectory to predict which part of the sky the object would be visible in. All that was required was for the relevant patch of sky to be checked out. If the asteroid was there, then the President and other world leaders would have to be roused from their beds; if not, we could all sleep more easily in ours. Unfortunately, however, there was a hitch. With cloud across much of the US, the large telescopes dedicated to asteroid spotting were unable to see anything, and it was left to another amateur, Brian Warner of Colorado, to issue the all clear, nine hours after the start of the imagined crisis. In the end, AL00667 proved—at 500 m—to be much larger than expected, but initial forecasts of its path were way out, with the asteroid approaching no closer than 12 million km—32 times further away than the Moon.

A number of lessons have been drawn from the events of mid-January 2004 that will have ramifications for how future potential threats from space are handled. In particular, the NEO Working Group of the International Astronomical Union recommends that, far from keeping the information secret, if any orbital solution for a newly discovered asteroid includes a possible short-term impact then the raw data must be made available immediately to the NEO scientific community, so that experts across the world can be involved in the interpretation and freely check each other's calculations. The group also advises that high priority should be given to involving as many observatories and observers as possible—both professional and amateur—in searching areas of sky predicted to contain a threatening asteroid should it be on a collision course. The task of eliminating such a *virtual impactor* must be given precedence over computing the correct trajectory, which would happen anyway within a few days.

Inevitably a time will come when we will be faced with a real, rather than a virtual, impactor. What happens next will depend upon a number of factors, the most critical of which are the time to the predicted impact and the size of the threatening object. An asteroid matching the original size estimate of AL00667 may be too small to be picked up in advance and the time available to do something about it could well be too short. On the plus side, however, the destructive footprint of an object in the 30–50 m size range is so small, compared to the area of the Earth's surface, that it is highly unlikely to score a direct hit on a town or city and it is much more probable that it will hit the sea. Bigger objects have the potential to trigger ocean-wide tsunamis in these circumstances, but a small asteroid will have only local effects. On top of this, recent research by Alan Harris of Boulder's Space Science Institute suggests that impacts of asteroids this big may occur only every thousand years or so, as opposed to every century or two, as previously thought. Many of these will also fail to penetrate the Earth's atmospheric shield, and will explode before they hit the ground. For asteroids 50 m or more across, such 'air-bursts' can still cause major local destruction, and the 60 m object that exploded 6–8 km above the Siberian wastes at Tunguska in 1908 flattened an area of forest equal to that of Greater London. The shock waves generated by asteroids even half this size, however, would scarcely reach the surface and would be unlikely to cause significant damage. The relatively low level of risk from such an explosion is summarized by Harris, who expresses the view that: 'I am about equally divided as to whether I would run away from the impact site or towards it.'

Much greater effort is being expended on ways of protecting our planet and our civilization from larger objects that are spotted early enough for us to attempt to do something about them. Some schemes are clearly more feasible than others, and the solution proposed by the Chinese in response to the discovery in 2003 that asteroid 2003QQ47 might strike the Earth on 21 March 2014 undoubtedly falls into the 'others' category. In an attempt to nudge our planet out of harm's way, all 1.5 billion citizens are being called upon to jump at the same time. Whether successful or not, it would certainly constitute a scientifically interesting experiment which, sadly, will now never happen—well, not in 2014 anyway. More accurate determinations of the orbit of 2003QQ47 have revealed that it has only a one in a million chance of striking the planet, so the great leap forward has been put on hold.

More seriously, however, NASA scientists and the US military, in particular, are getting itchy fingers, and plans are already afoot to test methods of directly intervening to change the orbit of an asteroid. As mentioned earlier, the old idea of blasting the object to bits with nuclear bombs or missiles—still much favoured by Hollywood—is a non-starter, and more subtle approaches are being seen as the way forward. To provide a measure of the task, it is worth appreciating that it only takes seven minutes or so for the Earth to move its entire diameter in space—a distance of some 12,672 km—so an approaching asteroid or comet only has to be given a very small nudge to ensure that a possible hit becomes a certain miss. Furthermore, the earlier or further away the nudge is applied, the smaller it needs to be, which is why the degree of advance warning is critical. One scheme being examined involves landing a rocket motor on the threatening object, which, when activated, would provide the thrust required to either slow down or speed up the mass of rock in its orbit so that it would miss the Earth when the encounter between the two eventually occurred.

At the 2003 British Association Festival of Science meeting in Salford, Matt Genge of London's Imperial College came up with a particularly populist way of demonstrating how this might work. Genge pointed out that the amount of thrust needed to provide sufficient shove was so small that it could be provided by Del Boy's clapped-out Robin Reliant car in the much-loved BBC television sitcom, *Only Fools and Horses*. This is a pretty extraordinary feat when you consider that the average 1 km asteroid tops the scales at around a billion tonnes and travels at close to 40,000 km an hour, while the humble three-wheeler weighs in at just 650 kg and takes a full 16 seconds to accelerate to 60 miles an hour (96 kph) from a standing start. The idea of a Robin Reliant's engine—which has a thrust just about comparable with the weight of an apple on the palm of the hand—or at least its high-tech, rocket-powered equivalent, being able to shift a mountain-sized mass of rock might seem like pie in the sky. However, it is worth remembering that in deep space, gravity is minimal and there is no air resistance; a small amount of force can start something moving and keep it moving, and an asteroid is no exception. In fact, as Genge points out, a velocity change of just 7 mm a second would be sufficient to ensure that a threatening asteroid missed the Earth. A motor pushing a billion-tonne asteroid with the force of a Reliant's engine would have an acceleration of just one billionth of a metre per second, but this would be quite enough to deflect the asteroid from its dangerous path in a little less than three months.

Of course, there are other factors to be taken into account, which would ensure that years would be needed to plan and undertake any mission aimed at diverting an asteroid, but the hope is that the current sky surveys will ensure that we have several years, if not—as in the case of aforementioned 1950DA—several centuries, to prepare. Such a mission, which might well have to be manned in order to cope with unforeseen problems, would have to rendezvous with the offending object, which is likely to be spinning slowly. Before any diversion could be accomplished the spinning would have to be stopped, something which, in itself, is likely to require the attachment of a precisely placed motor with an appropriate level of thrust. Once the asteroid has stopped rotating, the process of modifying its velocity can begin. The motor which would seem best suited to such a task is an *ion drive*, a rocket with an incredibly weak thrust but with the capacity to work for many months or even years provided there is sufficient fuel. An ion drive sounds like something from the realms of science fiction, and indeed it is—at least originally. In the 1968 episode of *Star Trek* entitled *Spock's Brain*, malicious aliens steal the Science Officer's brain—for whatever reason escapes me at the moment—and make off in an ion-powered craft. Given their characteristically low thrusts, it is not really surprising that the aliens were quickly caught by the Enterprise, Spock's brain recovered, and, if I remember rightly, popped back in his head by Leonard 'Bones' McCoy in a remarkable piece of computer-assisted surgery.

Ion drives have come along way since the 1960s, however, and are no longer confined to the scripts of space soaps. In fact, two ion-powered spacecraft are speeding away from Earth at this very minute. Despite encountering a number of problems on the way, NASA's *Deep Space 1* was launched in 1998, using ion power to hurtle past Comet Borrelly at almost 60,000 km an hour three years later. More recently, launched on 27 September 2003, the European Space Agency's ion-powered lunar probe, *Smart 1*, spiralled its way towards the moon, arriving in lunar orbit in November 2004—the most leisurely moon mission ever. Both spacecraft are propelled by the ejection of streams of charged atomic particles, rather than the chemical propellants used in normal rockets, which provide a tiny thrust about comparable to the weight of a postcard on the hand. Because the particles are travelling much faster than the hot gases blasted out by chemical rockets, however, ion drives are capable of accelerating spacecraft to very impressive velocities, given enough time. Chemical rockets simply blast out their fuel far too quickly to make this possible. Due to their efficiency, they also require very little fuel, and *Smart 1* made it to the Moon on just 82 kg of xenon propellant.

Attaching an ion drive or two to an asteroid that has the Earth in its sights is not the only potential means of providing the shove needed to bump it off course. An ingenious and lower-tech alternative involves changing the heat-conducting properties or reflectivity of the surface layers of an asteroid, so that the Sun's rays themselves can accomplish the required degree of deflection. This whole idea harks back to a classic paper written around 1900 by the Polish (or according to some sources, Russian) engineer, I. O. Yarkovsky, who proposed that solar radiation had a significant influence on the orbits of asteroids. Yarkovsky suggested that as the Sun warms an asteroid more on the 'day' side than the 'night' side, so the warmer side of the rock emits more thermal radiation. This results in what is now known as the *Yarkovsky Effect*, a difference in momentum that is tiny but significant, and which gives the asteroid a nudge. Depending upon the direction in which a particular asteroid is spinning, the effect can either speed it up or slow it down, sending it into an orbit further from or closer to the Sun. The effect may be at least partly responsible for sending asteroids Earthwards from the main asteroid belt between Mars and Jupiter and refreshing the population of Near Earth Asteroids. By modifying the way in which different parts of an Earth-threatening asteroid's surface heats up, it should be possible to utilize the Yarkovsky Effect to change its orbit in such a way as to ensure it misses its target. There are a plethora of different ways to do this, some of which seem distinctly silly. They are, however, being put forward seriously as a way of helping to protect our planet from asteroid impacts, and with no more enthusiasm than that shown by Joseph Spitale of the University of Arizona. He has come up with a number of suggestions for accomplishing the required changes in the surface thermal properties of an asteroid, which range from painting it or covering it in chalk or dirt to alter the manner in which it reflects sunlight, to using TNT to pulverize the surface layer so that the rate it conducts heat is modified.

Alternative ways of using the Sun's rays to modify an asteroid's orbit have been examined by Jay Melosh, another asteroid researcher at the University of Arizona. Melosh proposes attaching a giant solar sail to an Earth-threatening asteroid—an incredibly thin and light sheet of reflective foil kilometres across that is moved by the pressure of sunlight, taking the asteroid with it. Although elegant in principle, the solar sail solution might be difficult to pull off. Just like ocean-going yachts, the sail would need to have a sophisticated steering system in order to ensure that the asteroid was tugged in just the right direction, and it is likely to be difficult, if not impossible, to attach to an asteroid that is spinning—which nearly all are. An easier way of using a solar sail might be to 'collapse' one onto the surface of the asteroid, and leave the sunlight acting on the sail to provide a thrust directly to the rock. Melosh prefers yet another alternative method of using sunlight, this time by sending a giant mirror into orbit around the asteroid. The mirror would focus the Sun's rays on a pre-selected and carefully calculated part of the asteroid and vaporize or *ablate* the surface, sending material flying off into space. Like the recoil caused by a bullet leaving a rifle, the removal of this mass would push the asteroid in the opposite direction.

The ablation method might also allow the nuclear option back in. While blasting an asteroid into bits—which may actually be capable of causing up to ten times the damage of the original, undisrupted object—remains out of the question, nuclear explosives may still have a role to play in modifying the orbit of a bogie asteroid. The trick is to detonate the device several hundred metres above the surface of the object. This, it is thought, would be far enough to prevent the shock wave from breaking the rock apart, but close enough for the intense heat of the blast to ablate the surface material. A single stand-off nuclear blast, or more likely a series of blasts, could accomplish the required change in the object's orbit without having to land and attach engines or equip a team of astronauts with several million gallons of white gloss. First of all, of course, the nuclear explosives have to be launched into space—something that is likely to prove far from popular with those countries that lie beneath the launch trajectory.

NASA scientists are proposing to do away with the need for nuclear explosives, solar sails, and tins of paint, by proposing that an asteroid on a collision path is blasted by lasers. Not to blow it to bits, but, once again, to ablate off part of the rock's surface and use the resulting explosive release of gases to thrust the object off course. This method holds a number of advantages over attaching motors or a sail to an asteroid, in particular, doing away with all the hassle of matching speeds and landing on a spinning body. The ablation technique is also effective on a wide range of materials, so whether the target is composed primarily of ice, rock, or iron, it will still work. The approach is being considered by NASA scientists under the umbrella of the organization's Comet/Asteroid Protection System (CAPS) concept, an ambitious proposal aimed at developing an all-encompassing approach to protecting the Earth against asteroids and comets. The lasers could be based either in spacecraft or on the Moon, allowing incoming asteroids to be zapped remotely. Lasers powerful enough to do this remain on the drawing boards, but may well be in existence in coming decades, probably on the back of the US 'Star Wars' missile defence programme. They hold the advantage of delivering a strongly focused and very intense beam capable of pushing an asteroid from an enormous distance. Jonathan Campbell, a scientist at NASA's National Space Science and Technology Center, proposes that an array of solar-powered lasers could form an Earth Defence System based on the Moon. When and if an asteroid on a collision course is detected, the lasers would fire pulses at the object, at about the rate of a hundred a second, each burst ablating a little of the asteroid's surface and providing a tiny nudge. After weeks or months of continuous bombardment, the asteroid's course would be modified sufficiently so that it would pass by the Earth safely. Given powerful enough lasers, and six months' warning, a 1 km asteroid could be diverted in just a month, much less time than would be needed to prepare and launch a mission to the target asteroid.

8 The NASA Comet/Asteroid Protection System (CAPS) envisages installations on the lunar surface that use advanced multi-resolution optical/infrared telescopes to spot threatening asteroids and comets, and laser-ranging devices to get an accurate fix on the objects before pulsed lasers are brought to bear to divert the object into a safe orbit.

The proposed CAPS concept is about more than just zapping potential Earth-colliders, however, and its earliest goal is to learn more about the potential threat from space. In addition to continuing the search for large Near Earth Asteroids, CAPS incorporates a number of schemes to catalogue small NEAs below 100 m across—the potential city smashers—along with the *long-period* comets whose parabolic orbits bring them to the inner solar system, and our attention, only on rare occasions. The initial CAPS plan is to spot all large (1 km or more) Near Earth Objects out to a distance of five to seven *astronomical units* (an astronomical unit is the average distance from the Earth to the Sun, and amounts to about 150 million km), and all objects more than 50 m across within one astronomical unit of the Earth. Detecting long-period comets and quickly determining their orbits will form the real challenge, and success will depend on developing a detection system that operates beyond the Earth's hindering atmosphere. Advanced optical and infrared telescopes in orbit or on the lunar surface are proposed to hugely improve detection capability, with telescopes working in conjunction to develop better resolution and lasers being used as rangefinders to enable precise orbit determination. CAPS is not a programme that we can expect to reach fruition for many years, although preliminary funding is currently being requested by NASA. The current time-frame envisages detection and diversion capabilities being developed over the next 20 to 40 years, although it would not be surprising if this took far longer. Inevitably, politics will play a major role, and future administrations may not see the programme as a high priority. The pace of development is also likely to be dependent on how serious President Bush is about a return to the Moon, and whether or not this goal is taken up by future governments. Perhaps to make CAPS more attractive, NASA is also pushing the system's capacity to utilize asteroids as a resource, the idea being to manoeuvre objects into orbits where they can be reached and mined more easily. The mineral resource worth of even a 50 m asteroid, they point out, could amount to several billion US dollars. Personally, the most exciting aspect of the CAPS proposal seems to me the plan to solve the problem of a long-period comet on a collision course with our planet, although I must confess to some reservations about its chances of success. The CAPS solution, should we ever find ourselves in such a desperate situation—and have sufficient time to do something about it—is to initiate a game of interplanetary billiards: modifying the orbit of an asteroid so that it collides with the offending comet and pushes it off course.

Clearly, the capability of undertaking such a complex procedure remains some way off, but not as far off, perhaps, as the solutions devised by nanotechnologists. These promoters of the puny propose to counter the impact threat by unleashing legions of tiny nanobots. In his evangelical book, *Our Molecular World,* nano-enthusiast Doug Mulhall waxes passionate about just how the very small can be successfully utilized to overcome the very big. First, a million micro-satellites—each smaller than a thumbnail—would be scattered across the outer solar system in order to provide the ultimate asteroid and comet detection system. Closer to

Earth, a series of unmanned nanobot bases would be established in the asteroid belt between Mars and Jupiter. These would support the front line 'shock troops' in the battle to keep Earth safe: a fleet of special nanobots designed not to divert any threatening objects, but to eat them. Yes, *eat* them! Just a few microns across, billions of the little munchers would be carried in hibernation mode in a delivery pod powered by more bots combined to form a solar sail. On reaching the target asteroid or comet, the nanobots will revive themselves, replicate at an astonishing rate, and transform themselves into specialist extraction bots that will strip-mine the object layer by layer. Once the bots' stomachs are full, little would be left except a few bite-sized chunks of rock that would rapidly burn up in the Earth's atmosphere. Not only that, but by the time the object is due to rendezvous with our planet, Mulhall, envisages its being converted from a threat to a resource, having been transmogrified by the bots from a lump of rock into thousands of micro-factories manufacturing pretty much anything you like. Presumably, it is not completely impossible that part of an asteroid with your name on it could end up being a new lawnmower—with your name on it.

Meanwhile, back in the real world, serious plans are afoot to attempt to modify the orbit of an asteroid in little more than a decade, not because one is headed our way but just to see if we can. The plan is being touted by a group calling itself B612. While reminiscent of one of the long list of vitamins embellishing the side of a vitamin jar or a packet of breakfast cereal, the name derives from the legendary French children's novel *Le Petit Prince*, by the late pilot and poet, Antoine de Saint-Exupéry, and is the name of the asteroid on which the prince himself resided (awfully boring place apparently—absolutely no atmosphere). B612 is a California-based non-profit foundation that grew from a one-day workshop on deflecting asteroids, held at NASA's Johnson Space Center in 2001. Chairman of the Foundation's Board of Directors is ex-astronaut Russell (Rusty) Schweickart, who piloted the lunar excursion module during its test flight in Earth orbit during the *Apollo 9* mission in 1969. The stated aim of the foundation is to 'significantly alter the orbit of an asteroid in a controlled manner by 2015', which they hope to accomplish using a nuclear-powered rocket motor. This will push the selected asteroid rather like a tug boat, changing its orbit sufficiently for the modification to be observed from Earth, using radar. While not officially sanctioned by NASA, interest in the proposed mission is growing, and in April 2004, Schweickart testified about the project in front of the US Senate Sub-committee on Science, Technology, and Space. The key to the asteroid tugboat getting off the ground—literally—is liable to be NASA's recently announced *Project Prometheus*, the development of a nuclear-powered capability to support manned and unmanned missions beyond Earth's orbit. Critical to the success of Prometheus will be NASA and the US Government's skill in persuading a sceptical public that launching nuclear reactors over their heads is perfectly safe.

As well as considering methods of moving asteroids, ideas are now being tested about just how we should deal with specific threats from space. For a conference on planetary defence held in California in February 2004, David Lynch and Glenn Peterson of the Aerospace Corporation conjured up *defined threat* (DEFT) scenarios, which saw the Earth facing four approaching objects named after the musketeer heroes of Alexandre Dumas—D'Artagnon [sic], Porthos, Athos, and Aramis. D'Artagnon is a 150 m asteroid discovered in February 2004, which, for most of its orbit, is closer to the Sun than the Earth and is therefore often difficult to observe. When there are sufficient data to determine its orbit, however, it is realized that its next encounter with the Earth in 2009 will involve its crashing into Europe. This scenario was specifically put together to concentrate minds on what we might be able to do, in a hurry, and using today's technology.

Porthos is a much scarier proposition—a long-period comet discovered in February 2013 and due to strike Earth a little over two and a half years later in October 2015. With a diameter of 2 km, such a collision would be devastating, resulting in the deaths of maybe a quarter of the world's population, severe climate change, and the collapse of agriculture. Such a scenario would allow the application of technologies not yet developed or tested but would suffer from uncertainties not presented by asteroid D'Artagnon, most notably the fact that it would not be possible to absolutely constrain the comet's trajectory, due to the buffeting influence of jets of gas formed as the comet approaches the inner solar system and its ice starts to vaporize. This would mean not only that its impact site on the Earth could not be determined, but that it might not actually hit at all.

Athos is a 200 m asteroid spotted in 2005 and headed for a crash in the Pacific Ocean, a few hundred kilometres off the California coast, in 2016. To complicate matters, in 2009 it is discovered that Athos has a tiny moon of its own, a 70 m body called DeWinter. The Athos scenario provides sufficient time to get a better fix on its orbit and to launch exploratory missions to take a detailed look at the make-up and structure of the object prior to deciding the best way forward. It also poses the problem of what to do about two asteroids heading in our direction.

Finally, Aramis is another big monster—an asteroid close to 1.8 km across, discovered in 2006 and on target for an ocean impact 800 km off the Indian coast in 2033. Although large, the Aramis scenario provides plenty of time for missions to check out the object and consider the best approach to take.

While it would be nice to think that the conference delegates were not fazed by the DEFT scenarios, and put forward admirable and realistic plans to keep our planet safe, not until we are faced with such a threat in the real world can we have any idea of how we will react. I have no doubt that Tim Spahr, Reiner Ross, Alan Harris, and the others involved in the asteroid AL00667 débâcle have some idea of what it feels like to think that an impact is imminent, but the rest of us don't. Provided that we are not faced with a speeding comet on a collision course within the next few years, it is likely that we will soon have the capability to make a fair fist of defending our planet from attack. Having the capability is one thing, however, and actually using it quite another. In particular, lots of questions spring to mind with regard to international politics and how it is bound to get in the way of making a straightforward decision to square up to the invader from space. Who will decide? Who takes responsibility? Should the decision to go ahead require unanimity of the world's heads of state? Should the UN take the lead? Who will pay? Inevitably, cults and weirdos will pop out of the woodwork, embracing the asteroid or comet as God's revenge for the way we have lived our lives and/or ravaged the environment, and extolling a view that collision should not be prevented. Unquestionably, others will have managed to convince themselves that the object is in fact a giant spacecraft, cunningly disguised as a rock and populated by a race of altruistic super-beings come to save us from ourselves. Debate and discourse will inevitably rage about more serious issues. When should we decide to intervene to divert a threatening object? When there is a one in a thousand chance of impact? Or one in a hundred? Or one in ten? What if the international community is split over the need or worth of intervening? What if France says 'non' but the US says 'yup'. What happens if a nation takes a unilateral decision to go ahead with a diversionary attempt? Even more important, what happens if they get it wrong and translate a near miss into a certain collision? Somehow, my thoughts keep going back to the loss of that NASA Mars probe in 1999, because of a failure to convert imperial units into metric ones when calculating its trajectory. Maybe before launching any attempts to mess with the orbits of approaching asteroids or comets there should first be an international agreement that the metric system be used throughout. Certainly we should be considering the protocols and procedures that would come into play when and if a threat from space were detected. With perhaps just a few months to save our planet and our race, one thing we definitely don't want to get bogged down with is the drawing up and debate of long and complex wordings requiring signing and ratification by national governments. As they said of the rebuilding of astronaut Steve Austin in the cult TV classic, *The Six Million Dollar Man*, 'we have the technology.' The question is, on its own, will this be enough?

3

Tackling the Tectonic Threat

Many difficulties which nature throws in our way, may be smoothed
away by the exercise of intelligence.

Titus Livius (59 BC–AD 17).

Within a few minutes of 8 a.m. on a sunny Ascension
Day morning in 1902, the town of St Pierre on the
French Caribbean island of Martinique was transformed from a busy and bustling port into a charnel house of
blasted buildings and roasted human flesh. The bringer of
this terrible and near instantaneous metamorphosis was the
local Mont Pelée volcano, which, having rumbled ominously
for months, finally exploded with all the violence of a deton-
ating nuclear arsenal. No tardy lava flows here, no gentle fall
of volcanic dust; instead, a hurricane blast of incandescent ash
and gas vomited sideways and four-square at the unsuspecting
town and its doomed inhabitants. Contained within a valley
leading directly to the port, the ground-hugging black cloud
took just minutes to overwhelm St Pierre before scooting out
across the sea to bring death and destruction to many of the
ships anchored offshore. Although the volcano's rumblings
had been getting more and more violent in preceding weeks,
bringing ash fall and the smell of sulphur and rotten eggs to

the town, no one could have foreseen how events would unfold on that May morning. Indeed, no one on the island had ever seen or heard of a *pyroclastic flow*—the most deadly of all volcanic phenomena and the nemesis of one of the Caribbean's most handsome and cosmopolitan communities. Few of those caught in the blast were able to pass on their newly acquired experiences either, with just a handful of terribly burnt survivors living to tell the tale. The most famous— a prisoner called August Ciparis incarcerated in a massive-walled stone cell—went on to recount his amazing story and show off his scar tissue around the world, as part of the touring Barnum & Bailey Circus. Most of the 29,000 or so inhabitants, however, would never leave the island; their corpses burnt to a crisp, bloated and split by the heat, and trailing intestines and brain tissue from exploded chest and skull cavities.

Clearly, death by pyroclastic flow is a nasty experience and one to be avoided if at all possible. To a volcanologist, however, having the opportunity to watch and study such extraordinary phenomena close up is so unique as to make the risk of an untimely and unpleasant demise worth taking. Fortuitously, such an opportunity was provided in 1995, when the Soufriere Hills volcano on the British Caribbean island of Montserrat decided to end its three centuries of slumber and slowly but surely awoke. Colleagues from volcano-plagued countries like Italy, Iceland, the US, and elsewhere have always joshed the British geological community about its unusually fervent interest in volcanoes, given the fact that the last lava to flow in the UK did so in north-west Scotland some

40 million years ago. Now, at last, we had our very own active volcano, and British volcanologists swarmed to it like bees to a honey-pot. As one of those bees, a rather battered local airline turbo-prop dumped me unceremoniously onto the runway of the ramshackle W. H. Bramble airport early on a wonderful Caribbean evening in September 1996. For 14 months the volcano had been charging itself up and preparing for bigger and more violent things, and I was to do a stint as Senior Scientist at the Montserrat Volcano Observatory, to try to second guess what it had in mind next.

By the autumn of 1996, sticky and slow-moving lava had built a gigantic dome over the volcano's feeder conduit, a dome that periodically became unstable and collapsed, generating pyroclastic flows that poured down the eastern flank of the volcano and into the sea. Plymouth, the island's capital nestling at the western foot of the volcano, was by now officially evacuated, but persuading inhabitants to stay away from the exclusion zone that contained the southern half of the island was proving almost impossible. Against advice, many people remained in their homes in the village of Long Ground, close enough to the collapsing dome for the inhabitants to watch pyroclastic flows speed periodically past their homes. Just three days after I arrived, events unfolded that were to persuade most that this was not a good place to stay. As had happened on a number of occasions over the previous few months, the Soufriere Hills lava dome began to show signs of spontaneous collapse, starting late in the morning on 17 September. Pyroclastic flows surged down the Tar River valley (or *ghaut*, as the locals called these steep-sided, ephemeral watercourses) that drained the eastern flanks of the volcano, and surged across the sea just a few kilometres south of the still operating airport. Buoyed up by convection currents from the hot flows, great clouds of ash rose many kilometres into the sky before being blown westwards by the prevailing winds and plunging much of the island into stuffy, ashy blackness. Not wanting to miss the show, MVO scientists leapt into the five-seater chopper that supported our monitoring effort, and directed Canadian ex-Vietnam vet pilot, Jim McMahon, to head for the heart of the action. Hovering just a few hundred metres in front of a pyroclastic flow and its accompanying towering ash cloud is awesome in the true sense of the word—an experience that inevitably brings to mind thoughts of St Pierre and the fate of its inhabitants close to a century earlier. With the helicopter doors removed to provide better visibility, the radiated heat from the flows— even at this distance—is unbelievably intense, and it takes some effort to keep from checking exposed skin for the first signs of charring. So enthralling, so sensational is the experience that real fear never gets a look in. I would be lying, however, if I did not admit to some slight apprehension as Jim made the helicopter soar and wheel like a clumsy bird above and around the potentially lethal clouds.

By late afternoon things were quietening down considerably, and by 8 p.m. the lava dome had stopped collapsing. Those not on duty that night enjoyed a few beers, scoffed a couple of chicken rotis—the delicious meat-filled pancakes beloved of the islanders—and retired to their beds. Then, just before midnight and without warning, the volcano blasted back into life. As I poured a wee dram to hasten a good night's sleep, I became aware of a growing rumbling sound reminiscent of a jet engine starting up, soon to be accompanied by the howling of dogs and the crashing of thunder. I knew instantly what had happened; as we volcano-logists like to say the volcano had 'gone explosive'. In other words, the pressure exerted by gas dissolved in the magma had started to tear it apart, blasting it violently upwards and outwards. Rushing outside, I was greeted by a spectacular electric storm, as lightning bolts triggered by the eruption slashed across the sky. By the time I reached the villa that doubled up as the Montserrat Volcano Observatory, ash and pumice was beginning to fall and a great dark cloud slowly but surely ate up the stars. Red glows on the eastern horizon testified to new pyroclastic flows pouring down the volcano's eastern flanks, but in the darkness it was difficult to tell what else was happening. Just 6 km from the erupting vent, I doubt I was the only one with worries about survival. A major, sustained, explosive eruption now could send pyroclastic flows heading our way, bringing the prospect of a death too gruesome to think about. As small fragments of pumice con-tinued to clatter on roofs, people living closer to the volcano started to appear on foot and in their cars, moving north-wards to relative safety. To try to find out a little more about what the volcano was up to, I joined two others in a four-wheel drive and we took the coast road around the north end of the island in order to approach the eastern side of the volcano, where we would have a better view of things. The inevitable feeling of relief that accompanied the first part of the journey, which took us away from the fireworks, was replaced by one of apprehension as the road eventually took us around the top of the island and back towards the still erupting volcano on the other side. As we crept southwards again—ever on the lookout for the red glows of pyroclastic flows heading towards us—we gradually became aware that the lightning had stopped and the jet-engine rumblings had died down. Thankfully the eruption was over, after just 45 minutes. No lives were lost, but there were some close shaves. Flying in the helicopter low over the village of Long Ground the next day revealed a number of burning houses and many others with great holes in their roofs where chunks of volcanic rock travelling at terminal velocity had smashed through. At the height of the eruption an Air Canada jet passing over the island had flown into the column of ash, which had climbed rapidly to more than 10 km. Some fine ash insinuated its way into the cabin and some minor damage was sustained, but the plane continued on its way and eventually landed safely.

A mildly interesting account of an eruption, I hear you say. But so what? Well, our experiences on Montserrat and at other erupting volcanoes across the planet are crucial in that they help to clarify our picture of how volcanoes function, and in particular how they kill, maim, and destroy. Volcanologists are engaged in a constant struggle to improve our understanding of these lethal phenomena, often under very difficult circumstances, and over a dozen colleagues have lost their lives in the last 15 years while attempting to satisfy their thirst for more data. The more we know, however, the better placed we will be to cope with the wrath of a super-eruption when it

9 Pyroclastic flows are hurricane blasts of hot ash and incandescent volcanic gases that are totally destructive. Pyroclastic flows generated in 1997 during the eruption of the Soufriere Hills volcano on the Caribbean island of Montserrat obliterated the capital, Plymouth, and took 19 lives. Gigantic pyroclastic flows formed during the last Yellowstone super-eruption devastated an area of 10,000 square kilometres.

eventually arrives on the scene. Although the Soufriere Hills explosion is to a super-eruption what a fire-cracker is to the H-bomb, it does teach us a couple of very relevant lessons that we ignore at our peril. First, volcanic eruptions are often not predicted, not necessarily because we don't see the signs, but because we don't understand or interpret them as we should. Even as the 17 September eruption began to get going, most of us realized why it had happened. Remove the pressure exerted by overlying rock from gas-rich magma beneath the surface—as the collapse of the lava dome accomplished during the course of the previous day—and the gas will spontaneously form rapidly expanding bubbles that will tear the magma apart, blasting it upwards because this is the only direction it is free to take. Just like shaking a bottle of fizzy pop and unscrewing the top, the result is explosive and can be messy. The point is, on Montserrat we should have been ready for this. The collapse of the dome during the day should have led us to be aware of the possibility of a following explosive eruption, but it didn't. Retrospectively we understood, but in an emergency situation this is not good enough. If the blast had been larger and more sustained, many people—including ourselves—could have died, simply because we did not think through the possible implications of the day's events. Fortunately no harm was done, and later in the crisis—which, incidentally, is still going on—similar blasts were accurately predicted. Looking to the future, however, the big question is: will we be able successfully to predict the next super-eruption?

The second lesson the Montserrat blast teaches us is that

even small volcanic eruptions are incredibly powerful phenomena. While scoring just 3 on the Volcanic Explosivity
Index, the energy released by the 17 September explosion
was equivalent to a couple of Hiroshima nuclear bombs. Just
imagine then the power of a volcanic super-eruption a
hundred thousand times more violent, and you will begin to
get some idea of the kind of forces we will have to deal with in
tackling the threat of a cataclysmic volcanic blast. The rumblings of Montserrat's Soufriere Hills volcano may be pretty
small beer compared to the great super-eruptions of the past
at Yellowstone in Wyoming and Toba in Sumatra, but ultimately the processes involved in bringing molten magma to the
surface from deep within the Earth, where it is formed, are
just the same. These processes also provide tell-tale indica-
tors that fresh magma is on the move, with the result that no
volcano, large or small, erupts without warning signs.

Magma is generated in the *asthenosphere*, a relatively thin
layer near the top of the Earth's mantle—between about 30
and 200 km deep—where temperatures and pressures are
just right for rock to start to melt—a sort of 'Goldilocks'
region. As batches of magma accumulate, their low densities
relative to the surrounding solid rock cause them to rise
buoyantly, eventually taking them into the overlying *lithosphere*. This is our planet's brittle outer layer, made up of the
very uppermost mantle and the overlying crust. The great
tectonic plates that slowly dance across the face of the Earth
at about the speed fingernails grow are made up of the lithosphere, the dance itself taking place upon the partially molten and plastic asthenosphere. For magma to rise through

the brittle lithosphere it has to break the rock ahead of itself, a snapping process that triggers small earthquakes, which can be detected using networks of seismographs. Similarly, as magma approaches the surface and forms itself into reservoirs sufficiently large to feed eruptions, so it must make space for itself. Inevitably, as it does this, the ground surface above swells like a balloon, providing a sign that can easily be detected using modern monitoring technologies.

So no volcano should take us by surprise, provided, that is, that we are actually monitoring it, and for most of the world's active and potentially active volcanoes this is just not happening. No one knows exactly how many of the planet's volcanoes will erupt again, but the number is at least 1,500. This is how many have erupted at least once since the Ice Age ended 10,000 years ago, a very short time compared to the life spans of most volcanoes. It is not unusual for an individual volcano to be active for hundreds of thousands or even millions of years. Mount Etna in Sicily, for example, has taken half a million years to achieve its current enormous size, which entitles it to the prize for the largest continental volcano. Even more impressively, the gigantic oceanic volcanoes of the Hawaiian chain have taken many millions of years to reach their current stature, which sees Mauna Kea rising 10,000 m from the sea floor to form the world's tallest mountain, beating Everest by more than a thousand metres. Bearing in mind the longevity of volcanoes, it is highly likely that many that have not actually erupted in the last ten millennia still have the potential to do so. A better estimate of the total number of active and potentially active volcanoes is therefore probably around 3,000, but of these we are only monitoring around 150 or so, and many of these only superficially.

If we are ever to have a chance of managing a future super-eruption, it is absolutely critical that we are not caught napping. Advance warning of years, or even better, decades, could make all the difference in the world in terms of whether or not we are able to cope successfully with the event and its aftermath. To avoid a particularly unpleasant shock, therefore, we need to make a concerted effort to monitor more volcanoes more closely. Until this happens, it is perfectly possible that we could miss the warning signs that a gigantic volcanic eruption is on its way, simply because we are not looking. At this very moment, for example, a huge mass of magma that has been accumulating unnoticed for a century or more beneath the crust in a remote part of the southern Andes or beneath the jungle canopies of South-east Asia could be ready to blast its way to the surface, and we would know nothing about it until it was too late. Fortunately, new technologies are making volcano-watching that much easier, and Earth observation satellites are enabling scientists to monitor many volcanoes without actually having to visit them. This is a great boon for those countries, such as Papua New Guinea, that actually have more volcanoes than trained volcanologists.

Assuming that the magma ejected in a future super-eruption is only a part of the mass accumulated beneath the surface, then the total volume could be in excess of 10,000 cubic kilometres—enough to fill around two and a half million new Wembley stadiums. There is little doubt that the surface swelling associated with the emplacement of such an enormous mass of magma would be sufficiently large to spot from space, provided we were looking in the right place. But where should we be looking and what should we be looking for? Geographically, we can actually narrow down our search quite rapidly, as a super-eruption requires a recipe that is based upon ingredients found only in certain parts of the planet. Like all volcanic phenomena, the distribution of super-eruptions is not random, but is constrained by the pattern of tectonic plates that make up the Earth's exterior. Furthermore, these cataclysmic events require the sticky, rhyolite magma that is restricted to specific geological environments. In the sort of enormous volumes needed to drive a super-eruption, such a magma composition is found only at so-called *destructive plate margins,* where one plate is being driven down into the Earth's interior beneath another, or at appropriately named *mantle hot spots*—the same phenomena responsible for the huge flood basalt outpourings of the geological past—where a rising plume of hot mantle material is impinging upon and melting the overlying crust. Toba is an example of the former situation and Yellowstone the latter. With few exceptions, those geological environments capable of producing sufficient volumes of rhyolite magma are confined to the Pacific *Ring of Fire,* a zone of high volcanic and seismic activity that includes Alaska, the western US, the Kamchatka Peninsula of Russia, Central and South America, Japan and the Philippines, to South-east Asia and the South Pacific, and to New Zealand. Here then is where we need to focus our attention, but how can we narrow down the search further? There is absolutely no reason why a body of magma large enough to feed a super-eruption has to be associated with an obvious volcano that has erupted before, and it is perfectly possible for a new magma body to form where no volcano currently exists. If swelling of the crust associated with the accumulating mass of magma has been going on for thousands of years, its presence should be detectable by means of changes in the local topography, for example, modifications to the paths of streams and rivers. Depending upon the rate of tumescence, established rivers may either be diverted around the growing bulge or cut down into the rising topography forming ever-deepening gorges. If an existing drainage system is absent, one may eventually develop on the bulge itself as it gets higher and forms its own watershed, with streams and rivers draining radially off the swelling.

As well as being on the lookout for such new and potentially deadly volcanoes, we also need to keep a close eye on still active volcanic systems that we know about and that have spawned super-eruptions in the past. Such systems, like Yellowstone and Toba, bear the scars of previous great blasts in the form of giant craters known as *calderas*, which form during the later stages of eruptions when the undermined crust collapses into subterranean voids evacuated of their magma. At Yellowstone, three huge calderas up to 80 km across can be discerned, while at Toba the calderas formed by past eruptions have coalesced to form a single, enormous, lake-filled crater 100 km in length. Both the Yellowstone and Toba calderas are said by volcanologists to be *restless*, meaning that the crust at these locations is never still. Swarms of small earthquakes are common, and even the odd larger one, while the land surface may swell and subside by several metres across a huge area. Reminiscent of the slow breathing of a sleeping dragon, the most recent of the Yellowstone calderas, formed some 640,000 years ago, has swelled and subsided three times since the end of the last Ice Age. Most recently, after rising over a period of 50 years leading up to the mid-1980s, the caldera subsided during much of the 1990s before starting to swell again in 1999. This restlessness is a reflection of the fact that magma is still moving about deep down and may break through to the surface once again. More evidence for magma beneath restless calderas is provided by the fact that the amount of heat measured at the surface is far greater than in non-volcanic areas of the planet. At Yellowstone, for example, this so-called *heat flow* is 40 times the average for the Earth as a whole. It is not necessary to use any fancy gadgetry to confirm this because Yellowstone is dotted with mud pools, hot springs, and spectacular geysers, such as the mighty Old Faithful, which reveal the extent to which the subterranean magma heats circulating ground water and sends it back to the surface at temperatures easily high enough to boil an egg.

As mentioned earlier, we already have the capability to monitor volcanic unrest, both from space and on the ground. Because the methods used are likely to provide the first warning of a super-eruption precursor, let's take a look in a little more detail at how they work, starting with an ingenious system called *radar interferometry*. This utilizes satellite-based sensors to build a radar image of the area of interest on the Earth's surface and then constructs a computer-generated, three-dimensional model of the volcano, caldera, or other focus of attention. At some future time, weeks, months, or even years later, the same satellite takes a second image and produces a second model. With the aid of some high-powered electronic jiggery-pokery, the second model is superimposed on the first, with any differences between the two providing a measure of how much the topography has changed in the intervening period. The sensitivity of the method is really quite extraordinary and tiny movements as small as a few millimetres can be detected. The astonishing capability of the technique was demonstrated a few years ago by scientists at the innovative UK-based remote sensing company, Nigel Press Associates. Following the collapse of a block of flats in Rome during 1998, in which 27 people died, NPA researchers examined a series of satellite radar images showing the area in the weeks leading up to the tragedy. They were able to show that a few days before the collapse, the building started to deform. The movements were tiny—just a few millimetres—but they were the first signs that something was wrong and that the structure was becoming unstable.

The movements associated with a future super-eruption will undoubtedly be far larger, probably involving several metres or even tens of metres of uplift across an area of hundreds or even thousands of square kilometres. As we have not seen such a build-up we can't be certain of the scale, but we can get some idea by looking at swelling that has been observed at volcanic systems that have been active more

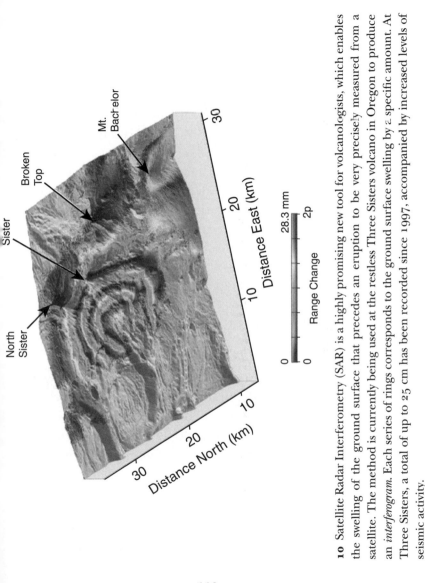

10 Satellite Radar Interferometry (SAR) is a highly promising new tool for volcanologists, which enables the swelling of the ground surface that precedes an eruption to be very precisely measured from a satellite. The method is currently being used at the restless Three Sisters volcano in Oregon to produce an *interferogram*. Each series of rings corresponds to the ground surface swelling by a specific amount. At Three Sisters, a total of up to 25 cm has been recorded since 1997, accompanied by increased levels of seismic activity.

recently. Like the Yellowstone caldera, the 15 km wide Campi Flegrei caldera, centred on the Italian town of Pozzuoli on the Bay of Naples, is restless. Unlike Yellowstone, however, the caldera has been active during historic times and last erupted in 1538. It has also been unusually agitated in recent decades. Twice in the last 35 years, in the early 1970s and again in the early 1980s, swarms of earthquakes accompanied rapid swelling of the caldera by up to 1.5 m, triggering widespread concern about a future eruption. No volcanic blast was forthcoming, but evidence from elsewhere suggests that an eruption could still start with very little warning. Similar bouts of seismic activity and ground swelling occurred during the mid-1980s at the Rabaul caldera in Papua New Guinea. Once again the situation quietened down after a couple of years, but then, in 1994, violent earthquakes and the sudden uplift of the ground surface by an unprecedented 10 m overnight provided just 27 hours' notice of a major explosive eruption that devastated the nearby city of Rabaul.

Satellite-based sensors will have little trouble spotting ground deformation on the scale seen at Campi Flegrei and Rabaul, and once they have identified a source of swelling that merits concern, the full panoply of ground-based volcano-monitoring techniques can be brought to bear in order to find out more about the form and size of the magma body beneath. The Global Positioning System is currently the favourite tool of volcanologists wishing to build up an accurate and detailed picture of ground deformation associated with the accumulation of magma beneath the surface. GPS is now commonplace, with car-based receivers becoming standard kit and hand-held models in use by sailors and walkers. Useful as they are, however, such systems are only good enough to allow users to position themselves on the earth's surface to within 10 m or so at best, and as such they are little use to volcanologists wanting to glean detailed information about a deforming volcano. The current US-owned GPS system is based upon a *constellation* of 24 satellites in Earth orbit that are continuously beaming radio signals down to the planet's surface. The orbits of the satellites are extremely precisely known, so locking on to four or more, using a GPS receiver, can give the exact position of the receiver on the Earth's surface to just a few millimetres—easily sufficient to detect the metre-scale swelling that we can expect to precede a super-eruption. By locating tripod-mounted receivers over benchmarks forming a network across a volcano, volcanologists can build up a picture of the pattern of deformation associated with the accumulation of new magma and, by undertaking repeated surveys, follow changes over time. They are also able to estimate, from the scale of deformation, just how much magma is down there, although not how much of this magma will actually come out during a future eruption. The current GPS system will soon be joined, in 2008, by a system called *Galileo*, developed by the EU and the European Space Agency, and comprising a constellation of 30 satellites. The system is designed to be compatible with both the US system and a Russian system called GLONASS.

Monitoring surface deformation on volcanoes is not a new technique and was in fact developed during the first few decades of the twentieth century by Japanese scientists studying the Sakurajima volcano. The same can be said for the monitoring of earthquakes at volcanoes, which began as early as the 1840s following construction of the world's first volcano observatory on Vesuvius. Seismic monitoring still forms the backbone of all volcano surveillance programmes because no volcano erupts without at least some precursory earthquake activity. Sometimes this is of such low magnitude that it can only be detected using seismometers. On other occasions, as at Montserrat, Campi Flegrei, and Rabaul, however, it can be strong enough to be felt or even to damage buildings. Volcanologists are able to learn a great deal from the seismic signals that are generated by accumulating magma. As a fresh body of molten rock rises towards the surface or an existing reservoir expands to accommodate a new influx of magma, so the overlying or surrounding rock is fractured. This cracking generates the sharp, seismic shock wave characteristic of a *tectonic* earthquake, which produces a distinctive signature on a seismograph. This is a strong signal with an abrupt start that tails off rapidly. The detection of many such signals at a volcano is usually taken as evidence of magma on the move, breaking rock time and again as it makes its way to the surface or builds a bigger reservoir prior to eruption. Numerous tectonic quakes are recorded during growth spurts of the lava dome at Montserrat's Soufriere Hills volcano and prior to each burst of activity at the Kilauea volcano's long-lived Pu'u 'O'o centre on Hawaii. Once magma has created a path for itself, the character of the seismic signals it generates changes. As a mixture of molten rock and gas passes through an open conduit or fracture, so it causes the walls to vibrate, rather like the walls of a house on a busy road when a heavy truck passes by. On a seismograph this results in a very different signal that does not have the strong and abrupt start of its tectonic equivalent and which lasts much longer, sometimes for many minutes. The appearance of such seismic signals, known as *volcanic tremors*, is a sure sign that magma is on its way, unhindered, to the surface, and that an eruption is likely within hours.

Accumulation of the huge volume of magma needed to feed a super-eruption will be accompanied by many years of tectonic earthquakes, some perhaps of a magnitude sufficiently large to inflict loss of life and serious damage to property across the affected region. Crunch time occurs once these give way to the volcanic tremors that herald the coming eruption, but is there any way of telling in advance how big this will be? A clue may be provided if it is possible to determine how much magma has accumulated beneath the surface. As mentioned earlier, the scale of ground swelling observed can provide some idea of the amount of magma available for eruption, but seismic studies hold the real key to building an accurate picture of the magma body. A technique known as *seismic tomography* is able to construct a three-dimensional model of a magma body within or beneath a volcano in the same way that a CAT scan builds a picture of the human brain, except it uses seismic waves rather than X-rays to achieve the final product. Because the properties of seismic waves change when they travel through magma as opposed to solid rock, by observing the behaviour of waves from remote earthquakes as they pass beneath the volcano of interest, it is possible to develop an image of the size and form of any magma body lurking there. This will provide an upper estimate of the volume of magma available for eruption and will at least be able to answer the question of whether a super-eruption is realistically on the cards or whether something smaller can be expected. Even if a tomographic survey reveals that a reservoir holds sufficient magma to qualify a future eruption for super-eruption status, it can't determine how much of the magma will be evacuated in the event. Great exposures of solidified granite, from the bleak moors of Devon and Cornwall to the towering peaks of the Andes, attest to the fact that much of the rhyolitic magma that feeds super-eruptions never makes it to the surface and instead cools and solidifies deep within the crust, eventually revealed by erosion. Consequently it is highly likely that we will never be able to tell beforehand if an impending eruption will indeed be worthy of the prefix 'super'—in other words, will have a volume of 1,000 cubic kilometres or more, or, if it does qualify, how much more magma than this will be evacuated. All we can do is prepare for the worst but hope for something considerably smaller.

If we can't say in advance exactly how big the next cataclysmic blast will be, we can at least have a decent go at predicting just when it will happen. Because no eruption occurs without warning signs, it is possible to study those signs to look for evidence of accelerating trends that might indicate that an eruption is on its way. Chris Kilburn at University College London has proved to be a particular dab hand at this, using seismic records retrospectively to predict eruptions at Pinatubo (Philippines) in 1991 and Montserrat's Soufriere Hills volcano in 1995. Although yet to be tested in advance of an eruption, it is likely that within the next few years improved monitoring will provide the data needed to make this possible. An impressive battery of surveillance methods is now at the disposal of volcanologists to support ground deformation and seismic monitoring, and to improve the chances of the successful prediction of an eruption. These include satellite sensors capable of detecting volcanic heat sources from space and techniques to observe changes in the rates of emitted volcanic gases, such as carbon dioxide and sulphur dioxide, which might warn of fresh magma approaching the surface. Other methods are capable of identifying and tracking variations in a volcano's gravitational, electrical, or magnetic fields due to the arrival of new magma. As technologies advance, monitoring instruments are certain to become more sophisticated. Already plans are afoot to develop intelligent surveillance networks made up of independent sensors able to communicate with one another. Imagine a situation in which a single sensor detects a tilting of the volcano's flanks that might arise from swelling, but which—before it sends out a warning message to the local observatory—checks with other sensors to make sure that they are observing the same phenomenon and that it is not the result of local movement or an electronic glitch.

The breathing space of days or weeks provided by eruption predictions is invaluable to civil authorities desperate to move their populations away from the threat of run-of-the-mill volcanic blasts. Although also critical for the evacuation of those living within the huge region likely to be devastated by a future super-eruption, this much warning is far from sufficient to allow the world to prepare for the climatic holocaust that may follow. Plans aimed at ensuring the survival of a functioning global economy and a coherent social fabric, and designed to support those countries least able to adapt to the post-eruption 'volcanic winter', must be far advanced by the time the seismographs warn of magma on the move. National governments must have in place food stocks sufficient for several years and a means of controlling their supply and distribution. Similar measures should have been taken by the time the first magma blasts its way through the crust to ensure that fuel stockpiles are sufficient to provide power and transport through the bitter weather. The length of the global freeze that followed the Toba super-eruption 73,500 years ago is thought to have lasted for about six years, followed by a cool period that lasted 1,000 years. Cambridge volcanologist Clive Oppenheimer has suggested recently, however, that the degree of global cooling might have been less than the 3–5 degrees C figure that is widely quoted. Should this be the case, the period of post-eruption cold might be neither as intense nor as long as expected. There is little doubt, though, that the next super-eruption will be devastating for the country or region in which it occurs, and globally disruptive. Provided the threat is recognized far enough in advance, and national and international preparations are immediate and sufficient, then the long-term impact on our society can be minimized. The response of the global community to contemporary climate change does not bode well for the sort of international cooperation that would be needed to manage the aftermath of a super-eruption, so perhaps we should lend an ear to some suggested ways of actually *preventing* such an event—at least for a bit of fun!

At the first inkling that a super-eruption is a possibility nuclear fixers will no doubt crawl out of the woodwork, enthusiastically advocating the use of hydrogen bombs to persuade the volcano to 'let off steam'. In the early 1990s an article appeared in the MIT-based newspaper *The Tech*, which proposed that neutron bombs could be an ideal solution to prevent lava flows from damaging settlements by blasting great holes in front of them within which the lava would accumulate. I can just see the residents of the pretty villages that dot the flanks of Mount Etna jumping at the chance of having their lovely mountain turned into the volcanic equivalent of a Swiss cheese. As an alternative to using brute force to short-circuit an impending super-eruption, nanobot devotees have suggested recourse to millions of grapefruit-sized robot excavators, whose task would be to burrow into the volcano to open pressure-relieving tunnels. The idea behind this is that when the volcanic blast finally happens, its strength has been much reduced, to the extent that hoards of *sky bots*, launched into the atmosphere immediately after the eruption, will be able to munch their way through the volcanic ash and gas, leaving behind them a sparkly clean atmosphere and circumventing any possibility of a crippling volcanic winter.

Far-out schemes involving nuclear ding-dongs and burrowing holes into magma reservoirs completely ignore the unimaginable energies locked up in these overwhelming natural phenomena. A volcanic super-eruption might well burn up as much energy as the entire US uses in a year, while a giant tsunami crashing onto the east coast of North America will expend as much energy on *every* 100-m stretch of shore-line as the World Trade Center buildings generated when they collapsed. At best, exploding nukes and burrowing bots will have about the same effect on a giant magma reservoir as a mosquito probing an elephant's hide, at worst, they might just provide that final, tiny shove that prompts the magma to blast its way to the surface ahead of time. Science and tech-nology are never going to prevent the next super-eruption, but they will help us to cope with its aftermath by spotting the threat in advance, giving us some idea of its potential size, and providing some warning—albeit a short one—of when it is likely to happen. Armed with this information, there can be no excuse for the international community failing to devise contingency plans designed to minimize its impact. All that is needed is for nations to unite to face a global geo-physical threat with the potential to affect everyone on the planet. Now, why does that not fill me with confidence?

To our, admittedly incomplete, knowledge, no giant magma reservoir is currently primed and ready to go. The same cannot be said, however, for the gargantuan landslide that rests rather uncomfortably on the western flank of La Palma's Cumbre Vieja volcano. As discussed in Chapter 1, this colossal mass of rock threatens to plunge into the adjacent ocean, generating catastrophic tsunami capable of unprecedented destruction and loss of life around the entire

North Atlantic rim. The Boxing Day 2004 tsunami graphically demonstrated how these lethal waves can travel across entire ocean basins in a matter of hours, transmitting death and annihilation to locations far from the source. While the Indian Ocean tsunami were rarely more than 10 m high, however, computer models predict that waves exceeding 100 m in height will be the awful legacy of the future collapse of the Cumbre Vieja volcano. Can we forecast when this is going to happen? Is there anything we can do to stop it? How can we manage or mitigate its awful consequences for the inhabitants of La Palma, the neighbouring Canary Islands, and all those great cities clustered along the Atlantic's seaboards? Forecasting collapse of the landslide is tied intimately to predicting future eruptions of the volcano. This is because the landslide will almost certainly be shrugged off during the turmoil of an eruption rather than in the intervening quiet episodes when magma is lying low. Our monitoring during the 1990s suggested that the slide was probably creeping seawards at perhaps a centimetre a year. This rate of movement could reasonably be expected to increase significantly when fresh magma is emplaced within the volcano, triggering an eruption. In fact, there has already been one eruption since the slide was initiated in 1949, with magma reaching the surface in 1971 right at the southern tip of the island. Whether or not this resulted in any acceleration of the detached mass we simply don't know, as no one was looking at the time. It is likely, however, that this particular event occurred far out of harm's way and what we really need to worry about is a future eruption that happens high on the flanks of the volcano adjacent to the slide itself. Not only might magma feeding such an eruption provide a mechanical push that could help to get the slide moving, but shaking caused by earthquakes linked to the eruption might also be expected to encourage the landslide to begin its journey to the seabed. Furthermore, the heat of the newly arrived magma within the volcano might also prove effective at increasing the pressure of groundwater contained in rock, causing it to exert the extra push that could provide the final straw.

The Cumbre Vieja volcano is one of the most active in the Canary Islands, and over the past 500 years it has erupted six times. These eruptions are not equally spaced in time, however, with periods of dormancy lasting from 22 to 237 years. We may therefore have to wait until 2171 or later before magma once again blasts its way to the surface on La Palma, and even then we have no idea if this will trigger the landslide. Indeed, no one has the faintest idea whether it will take five, ten, twenty eruptions—or more—finally to send this great rock mass to a watery grave. In order to have any chance of predicting when the collapse will occur, it is vital that an improved surveillance network is in place, ideally prior to the next eruption. At present the three existing seismographs are capable of providing an early warning of fresh magma entering the volcano, but little else. No instrumentation is yet operational that can build a picture of the deformation that will also accompany the rise of fresh magma and which may herald rejuvenation of the landslide itself. The technology is certainly available to do this, and the establishment of a regularly surveyed GPS network could provide all the information needed to help predict not only a future eruption but also increasing instability and accelerating movement of the landslide. Similarly, satellite radar interferometry could provide regular annual check-ups on the landslide's behaviour, not only into the future but also in the recent past. Plans are afoot to use the method to compare over seventy images of the volcano captured from space during the last decade and a half, in order to confirm whether the slide has been on the move over this period. This, it is hoped, will provide a much better picture of what the slide has been doing recently than the brief snapshot provided by the curtailed ground deformation surveys undertaken by my team during the 1990s. Crucially, it might also help to delineate better the size of the moving mass, which is at present rather poorly defined. The greater the volume of material that enters the ocean without breaking up, the bigger the resulting tsunami and the more devastating the effects when the waves strike the coastlines of those countries bordering the Atlantic. Current estimates vary between 150 and 500 cubic kilometres, with the lower end of the range predicted to generate 3–8-m high waves arriving on the coasts of the Americas, and the top end battering the same coastlines with waves up to 25 m high. Clearly, knowing which scenario is more likely would help considerably the national governments and their emergency authorities charged with coping with the event and its aftermath.

We have then the capability to monitor the Cumbre Vieja landslide and to predict the next eruption of the volcano, probably a week or two ahead if not longer. We also have the technology to get a better idea of the true size of the beast and therefore its likely impact, via tsunamis, on the Atlantic rim. I will return later to the measures we will have to take once the situation appears critical, but in the meantime, is there anything we can do to actually prevent the collapse occurring? Well, once the landslide really starts to shift, nothing can stop it. The moving mass is likely to be shaped a little like a half-submerged wedge of cheddar cheese lying on its side with the thin half under water. The maximum dimensions are estimated at 25 km long, 20 km wide, and perhaps 1–2 km thick, and the whole chunk will probably move as a coherent block for perhaps 15 km before breaking up and cascading down to the ocean floor some 4 km down. Despite its enormous size, the slide will be no slouch, hurtling off the side of the Cumbre Vieja at up to 360 km an hour—about as fast as Michael Schumacher's Ferrari. If we can't stop the slide once it is on the move, can we do anything about it beforehand? I am sometimes asked, for example, if we can quarry out the unstable mass and remove it bit by bit. This might seem a reasonable suggestion, but there are a number of problems with it, not least the enormous mass that would have to be removed. Assuming that an open truck can carry away 10 cubic metres of rock at a time, this would require between 15 and 50 billion journeys to take it away. Even with a loaded truck leaving every minute of the day, this would take somewhere between 10 and 35 *million* years. And this does not even take into account the time needed to excavate the material, nor the fact that a good percentage of the slide is under water. No doubt the 'nuke 'em' cheer leaders will enthusiastically put forward their usual solution, but just who would be foolish enough to detonate nuclear explosives on an already critically unstable giant landslide? I am always struck by some people's enthusiasm for the nuclear weapon as an all-purpose, solve-everything device, whether it be blowing apart asteroids, short-circuiting volcanic super-eruptions or blasting unstable islands. A few years ago I was telephoned by a very well-known writer of maritime thriller novels, who wanted to know if the Cumbre Vieja landslide could be triggered by nuclear missiles. These books usually have a big nuclear submarine or surface warship on the cover and occupy pride of place in airport bookshops. Apparently, the author was planning a new work in which terrorists captured a US nuclear missile submarine and then blackmailed the government by threatening to launch the missiles at La Palma, triggering collapse and devastating the east coast of the US due to the resulting tsunami. The book—*Scimitar SL2*—hit the UK bookshops in July 2004 and stars a not very well disguised me as Paul 'Lava' Landon. Sadly, having been pumped for information by Hamas terrorists, I don't last long and my bullet-ridden body is found somewhere around page 20. Let's hope that no one in the real world gets any ideas. Notwithstanding whether or not the plan would ever work, it has since struck me just what a ludicrously flawed idea this really is. If terrorists ever managed to get hold of a submarine loaded with nuclear missiles—each with multiple warheads—wouldn't they just launch them directly at the US? There would certainly be enough to wipe out most major cities in the country, including those on the west coast and in the interior, which the waves from La Palma could never touch.

Assuming that we can't prevent collapse of the western flank of the Cumbre Vieja, how can we minimize the impact of the event when it happens? Advance warning is going to be critical if the death toll resulting from the tsunami is not to be unimaginable, particularly in the Canary Islands themselves, in the Caribbean, and in the packed cities of the eastern seaboard of North America. With the waves predicted to sweep clean the coasts of the Canary Islands within 60 minutes of collapse, and take just 6 to 12 hours to cross the Atlantic, there will be little time for flight once events have begun. Only the greatest mass evacuation in history can prevent enormous loss of life, with perhaps in excess of 50 million people needing to be moved to safety along the east coast of the US alone. The populations of the many low-lying islands of the Caribbean will require temporary relocation to North or South America or Europe, while coastal communities in northern Brazil and north-west Africa, which will also face devastating waves, would need to be moved inland.

All well and good, but when to take the decision to move threatened people out? When an eruption starts? When the landslide starts to show signs of accelerating? By then, unfortunately, it may well be too late. Collapse of the north flank of Washington State's Mount St Helens volcano, which triggered the much-filmed 1980 blast, provides a useful example of what we could expect at La Palma. At the US volcano fresh magma, unable to get out of the blocked summit vent, intruded itself into the side of the volcano, causing it to bulge outwards. The swelling grew at pretty much the same rate for over a month, with monitoring scientists keeping a careful lookout for any acceleration of the swelling, which they expected to precede a collapse of the increasingly unstable mass. Then, on 18 May and without any warning, the entire north flank collapsed spontaneously, killing 57 people who had ignored advice to leave the area, and leading to a huge explosive eruption. No acceleration in the rate of growth of the swelling had been observed prior to the collapse; instead, a moderate earthquake beneath the volcano had literally shaken the bulge until it simply fell off. If the Cumbre Vieja landslide is reactivated during a future eruption of the volcano, then it is perfectly possible that the same sequence of events could occur. This means that there may be no warning of the precise time of collapse and that arrangements for evacuation of the areas at risk must begin at least as soon as there is any movement of the slide. Undoubtedly, this is going to put heads of state in incredibly difficult positions. Should they move their people out on the off chance that small movements of the landslide will be translated into the feared wholesale collapse? Or should they hang on and hope that nothing further happens? What if they make the wrong call? If a future US president initiates a mass evacuation from Boston, New York, Miami, and countless other coastal towns and cities under threat, and nothing happens, his chances of political survival are likely to be slim indeed. Furthermore, few people would be likely to respond to a future evacuation call when the volcano next awakened. On the other hand, to do nothing would be to court the possibility of losing a substantial portion of the country's inhabitants.

Notwithstanding such thorny issues, there are things we can do now to help mitigate the event when it eventually happens. We can start ensuring that some form of contingency evacuation plan is in place in order to speed up its implementation should the need arise. We can also argue for effective tsunami early-warning systems in the Atlantic Ocean so that even if evacuation is not completed by the time of collapse, at least some lives can be saved in the several hours' breathing space available to communities remote from the Canary Islands. Because of the large numbers of earthquake-triggered tsunami, which have taken more than 50,000 lives in the twentieth century alone, the Pacific already has a tsunami early-warning system, established back in 1948. Over 30 seismic monitoring stations located in and around the Pacific relay information about earthquakes in the region back to the HQ in Honolulu, Hawaii. If a quake larger than Richter Magnitude 7 occurs, tide-monitoring stations close to the epicentre watch for anomalous wave activity that might suggest that a tsunami has been triggered. If this is detected, a tsunami alert is issued that should reach all coastal communities within an hour of the quake. With hindsight we know now that if such a system had been in place in the Indian Ocean prior to Boxing Day 2004, perhaps a hundred thousand or more lives could have been saved in Thailand, Sri Lanka and southern India, which the tsunami from the Indonesian earthquake took two hours or so to reach. A similar system in the Atlantic would provide mainland Spain and the UK with two or three hours' notice, and the Caribbean islands and eastern seaboard with five to eight hours' warning, sufficient to save considerable numbers of lives. Such an alert will do little to help residents and holiday-makers on the Canary Islands themselves, where the waves will arrive just 30 minutes to an hour after collapse. A regional tsunami alert network, perhaps linked to a civil defence alarm system, could still, however, help to reduce the death toll. Such a system might be modelled on that developed in 1952 by the Japanese Meteorological Agency, which seeks to issue a tsunami alert within 20 minutes of a tsunami-generating submarine quake within 600 km of the country's coastline. During the 60 years leading up to the launch of the alert system over 4,000 lives were lost due to tsunamis hitting Japan's vulnerable east coast. Since then, however, tsunamis have claimed just a few hundred lives.

While timely evacuation may save tens of millions of lives, it can do nothing to reduce the exposure of property to the giant waves. Particularly in areas like Florida, the Bahamas, and other low-lying area, the tsunami may travel several kilometres inland, obliterating most man-made structures. Measures to reduce wave impacts are adopted in areas regularly at risk from earthquake-generated tsunamis, such as Hawaii and Japan, including barrier walls and forests to arrest and break up the waves. Following the collapse of the Cumbre Vieja, such measures could prove effective along the southern coasts of the UK and Spain, where wave heights are liable to be less than 10 m. No government or local authority, however, is going to fund an anti-tsunami defence programme against an event that may not occur for a thousand years or more. If we do have to wait this long, or longer, then the nanotech evangelists may have an answer—coastal tsunami barriers taking the form of graphite curtains reinforced with nanotubes. These would lower themselves from floating offshore barriers, unravelling to form a sloping, artificial shoreline that becomes rigid as an electronic command changes the curtains' molecular structure. The waves simply batter themselves into oblivion against the ersatz shore until they run out of energy. Nothing could be simpler. If and when we have the technology to accomplish such fantastic micro-feats, however, the cost of protecting the Atlantic rim would still appear to be astronomical. Furthermore, the Atlantic is not the only ocean basin threatened by giant tsunami, so even this mind-boggling cost is likely to be multiplied tenfold.

As discussed in Chapter 1, giant tsunami-generating landslides have been recorded across the planet, resulting both from collapsing ocean-island volcanoes and the steep margins of the continents in all the major ocean basins. While La Palma seems to constitute the greatest contemporary threat, it is not the only one. For example, one of the volcanoes in the Cape Verde chain—off the coast of West Africa—is also showing signs of becoming unstable. We also have the climate change dimension to consider, with evidence that rising sea levels and wetter conditions may promote more collapses of island volcanoes. In addition, concerns are also being voiced that warming of the world's oceans will lead to the destabilization of enormous deposits of solidified methane gas—known as *gas hydrates*—stored in marine sediments. The breakdown of such deposits into gas and water has been implicated in triggering the Storegga Slide that sent a tsunami surging across north-east Scotland 8,000 years ago. Future giant collapses of sediment could occur at any of the continental margins as climate change begins to take a tighter hold over the next hundred years and beyond. The threat of giant tsunami therefore stretches far beyond La Palma, but should we really be considering defending all the world's shorelines from these watery terrors? Clearly this is totally unrealistic and a solution that could never be sold to the international community. Giant tsunami undoubtedly have the potential to be a major global problem, but like super-eruptions, prevention is not an option. The best we can hope for is to mitigate and manage these watery terrors through a combination of raised awareness, better monitoring, improved communication about the threat, and forward planning.

Of all the terrestrial gee-gees, catastrophic earthquakes are unique in that we are in a position dramatically to limit the physical damage they cause as well as reducing death and injury. No earthquake has ever been or ever will be great enough to impinge physically upon everyone on the planet. The size of an earthquake is limited by the strength of the Earth's crust, which can only cope with so much stress before it breaks. The energy released when this occurs is tremendous, and for the biggest quakes may be the equivalent of tens of thousands of Hiroshima-sized nuclear blasts, but the vast bulk of the energy is quickly dissipated and the severe shaking that causes destruction and loss of life rarely extends beyond a thousand kilometres. Nothing recorded in historical times has exceeded 8.9 on the Richter Scale of earthquake magnitude, and this seems to pretty much represent an upper limit for earthquake size. As usual, however, this has not prevented US film-makers from overhyping the seismic threat with the launch in 2004 of *10.5*—a very high profile TV mini-series that had US seismologists tearing their hair out in frustration. The film, which sees a gigantic quake—scoring a magnitude of 10.5—separating western California from the rest of the US, has been well and truly hammered by scientists and state officials whose job it is to manage responsibly the earthquake threat in California. As the United States Geological Survey has been at pains to point out, earthquake size is limited not only by the strength of the crust but also by the length of the fault that ruptures to produce it. A quake of magnitude 10.5 would need a fault to rupture that is long enough to encircle the Earth, and such a fault does not exist. Furthermore, the longest rupture ever recorded in California—triggering the quake that devastated San Francisco in 1906—is just 400 km long, more than 39,000 km shorter than that required to generate a magnitude 10.5 earthquake.

An earthquake can be far smaller than that portrayed in the latest Hollywood fantasy and still wreak global havoc, provided that it strikes at the right time and in the right place—or perhaps that should be the wrong time and the wrong place. The havoc won't arise from the quake's physical impact but from the economic shock waves it transmits across the world almost as fast as their seismic counterparts. While a number of cities in the industrialized countries face the threat of a major earthquake, including Los Angeles, San Francisco, and Seattle in the US, the destruction of only one city has the potential to initiate a global economic meltdown to rival if not exceed that of 1929, when the Wall Street Crash resulted in the closure of 100,000 companies in the US alone. As mentioned in Chapter 1, the Tokyo–Yokohama megatropolis contains over 33 million people and is the heart of the Japanese economy. With the estimated maximum cost of the next 'big one' to strike the region set at 4.3 trillion US dollars—over three times the combined cost of *all* the natural disasters occurring over the last 50 years—many in the risk business await the catastrophe with trepidation, fully aware that the collapse of a weak Japanese economy could have devastating ramifications across the globe.

We can no more prevent an earthquake (but more later about trying to reduce their size) than we can snuff out a super-eruption or neutralize a giant tsunami, but we can greatly diminish its impact on people and property. An earthquake cannot be predicted in the same way as a volcanic eruption, because there are not always warning signs—or at least signs that are consistent. Like a volcano, however, a so-called s*eismogenic* fault that triggers earthquakes is only occasionally a threat. For much of its life a fault will quietly and unobtrusively accumulate strain, only to release it periodically in the short, sharp bursts that are earthquakes. Every active fault will generate quakes according to a uniquely characteristic timetable. Like buses and trains, however, earthquakes don't stick rigidly to this timetable, so their arrival can't be accurately forecast. Nevertheless, an average *return period* for a fault can be worked out by looking at the frequency with which it has produced quakes in the past. This can at least provide some clue to when the next one might be expected. Because of the somewhat vague nature of these forecasts, they are usually presented in terms of probability. For example, the United States Geological Survey reports that there is a 62 per cent probability of at least one magnitude 6.7 or greater earthquake in the San Francisco area before 2032. For southern California the forecast is for a more imminent event, with an 85 per cent chance of a magnitude 7 or greater earthquake within the next three decades.

No serious earthquakes have troubled the Tokyo–Yokohama region since the cataclysmic Great Kanto Earthquake in 1923, but the region is now at risk from quakes occurring on no fewer than four local faults. This part of Japan is geologically incredibly complex, with four of the Earth's tectonic plates converging here. The result makes the return periods of earthquakes on individual faults even more difficult to forecast because the movement of one fault may trigger a quake on another. A magnitude 6.5–7 earthquake is thought to be imminent, however, about 75 km south-west of Tokyo, close to the city of Odawara. Although unlikely to be devastating, such an event can be expected to cause moderate damage in the Greater Tokyo Metropolitan Region. A so-called *Tokai* quake is also due, beneath Suruga Bay 150 km to the south-west. Here, a much larger event, possibly registering magnitude 8 or more on the Richter Scale, may have a 35–45 per cent of chance of happening by 2010. Once again, the capital is unlikely to suffer serious damage, although the coastal city Shizuoka could be badly hit. The two greatest threats to the region come from what is termed a *chokka-gata* quake directly beneath Tokyo and a repeat of the Great Kanto Earthquake beneath Sagami Bay to the south. A *chokka-gata* event, perhaps as strong as magnitude 7, is thought to be on the cards and would lead to severe damage to the capital. A repeat performance of the magnitude 8.2 earthquake that struck in 1923 is, however, the true 'big one' whose potential global consequences—when it happens—may be severe enough for it to merit gee-gee status. Compared to when we are likely to face a future super-eruption, a giant tsunami, or an asteroid impact, the next Great Kanto quake is just around the corner. As the gap between the last two earthquakes—in 1703 and 1923—amounted to 220 years, however, we may have a while to wait yet.

Forecasting roughly when a future quake might occur is all well and good, but it is unlikely to save many lives. To do this, accurate prediction is needed, ideally far enough in advance that people most at risk can be evacuated, power stations shut down, mains gas and water supplies switched off, data saved, the emergency services prepared, and other essential measures taken. Unfortunately, no one has yet convincingly predicted an earthquake, although a number of individuals and research groups have made this claim. Japanese scientists still hold great faith in being able to spot warning signs before the next 'big one' strikes the capital, and there are indeed certain phenomena that are often observed before a big shock. Groundwater may migrate into new cracks opened by

11 Following the Great Kanto Earthquake of 1923, up to 40,000 survivors congregated in the open space of the Honjo Military Clothing Depot in an attempt to escape the huge firestorms rampaging through the ruins. There was to be no escape, however, and all but a few hundred were immolated as the walls of flame over-ran the temporary refuge.

the enormous strains that have accumulated along a fault prior to its rupturing. This in turn may lead to a fall in the level of the water-table, which can be detected by monitoring levels in specially dug wells. Detectors in the same wells can also check for levels of the radioactive gas *radon*, which is able to pass up through the newly opened cracks from deep within the crust. Studies from around the world of water-table changes before earthquakes have shown that, broadly speaking, the greater the area affected and the longer in advance the changes are observed, the bigger the quake will be. Seismologists have also hijacked many of the techniques used by volcanologists to look for the signs of surface deformation that indicate magma on the move. The Global Positioning System, in particular, is now in common use in areas of high seismic risk across the planet, precisely recording tiny movements of the ground surface close to seismogenic faults in the hope of discerning accelerating displacements that might indicate that a quake is on its way. A big quake may also be signalled by an advance guard of moderate tremors, known as *foreshocks*, and this appears to have been the case before the great San Francisco earthquake of 1906. Tiny quakes known as *microseisms* are also sometimes recorded in the run-up to something far bigger, probably arising from the formation of cracks along that part of the fault that is ready to snap. You might think that with all this going on beforehand it must be almost impossible *not* to be able to tell when a major earthquake is on its way. Unfortunately, however, not a single one of these phenomena can be relied upon. You *may* see changes in the water-table or in levels of radon gas seeping from the crust before an earthquake, or you may not. Increasing displacements across a fault may indicate that it is building to tear itself apart, but not necessarily. A sequence of smaller quakes may herald the 'big one' or they may simply occur in isolation. In other words, none of these potentially useful predictive phenomena is observed *every* time before every quake, so they are just too unreliable to use. It is unrealistic to expect the population of an entire city to evacuate because there *might* be a big earthquake. The predicted event has to be certain or, at the very least, its occurrence has to have a very high level of probability.

No stone is being left unturned in the search for the true Holy Grail of geophysics, the ability to predict accurately and reliably potentially damaging earthquakes, and even the animal kingdom has been drafted in. Like many others working in the hazard and disaster business, I get my fair share of loony letters. In particular, my mind goes back a decade or so to an especially bizarre one from a woman in London, who claimed that her cat had started behaving eccentrically just before Saddam Hussein's invasion of Kuwait. I remain to be convinced of the abilities of cockney cats remotely to predict small Middle Eastern conflicts, but reports of animals behaving unusually before earthquakes have been coming in for many years from all over the world, and they may have some basis in fact. In Japan itself fisherman have claimed bumper catches in the run-up to a major quake, while the much revered Japanese catfish is said to become particularly excitable before a major bout of ground shaking. This is somewhat appropriate as the Japanese have traditionally explained away earthquakes as being due to the squirmings of a subterranean giant catfish that goes by the name of Namazu. All sorts of animals, from chickens to pigs, have in fact been charged with adopting strange behaviour in advance of a big seismic event, perhaps due to electrostatic charges building up in fur, feathers, or scales. The point is, can this 'strangeness' be objectively measured and is it displayed before every quake? The answer is probably no, and although one group of Japanese scientists are busy scrutinizing a bunch of laboratory catfish in the hope of saving the capital from destruction, I feel that they are likely to have a long and fruitless wait.

The catfish might not be able to come up with the goods, but all is not yet lost in the quest for the Grail. A number of research groups have maintained that they are able to predict earthquakes accurately, using geophysical data rather than woolly accounts of niggly pets and boisterous farm animals to back up their claims. Most notable are the Greek trio who go by the collective name of VAN (after their surnames—Varotsos, Alexopoulos, and Nomicas), who announced as long ago as 1981 that they had spotted so-called *seismic electrical signals* (SES) before earthquakes in their native country. The team claims to have forecast a number of Greek earthquakes, including the one that struck Athens in 1999, but many seismologists remain unconvinced. The problem is that Greece is one of the most seismically active countries in the world, so any prediction of a quake months ahead is bound to be fulfilled. It may be, however, that the VAN team is on to something, claiming that their mysterious signals are formed as a consequence of stress changes in the Earth's crust before an earthquake. There is evidence that electrical signals can be produced by rocks under great stress, and a few years ago the physicist Friedemann Freund, of California's San Jose State University, proposed that these might in fact be able to generate huge electrical fields at the surface, causing those enigmatic phenomena known as *earthquake lights*. These have been reported before many major earthquakes and even filmed and photographed. As long ago as 1968, Yutaka Yasui, of Japan's Kakioka Magnetic Observatory, captured images of red streaks in the sky and a blue glow close to the surface, while even further back, luminous fireballs and glowing lights were reported before the great Lisbon quake of 1755. Lights in the sky were spotted before a quake of magnitude 7.8 obliterated the Chinese city of Tangshan in 1976 and were filmed by Turkish television before the devastating Izmit earthquake in 1999. Freund claims that earthquake lights are actually glowing plasma formed when the powerful electrical fields generated by extreme strain in the crust ionize the atmosphere. Energy is also released from the stressed rocks in the form of infrared radiation, which we can't see but—crucially—satellite sensors can, offering the possibility of predicting future earthquakes from space. NASA remote sensing specialists, Nevin Bryant and his team, have taken things a step further by checking satellite records for signs of infrared emissions before the Izmit quake and the devastating earthquake that shook the Bhuj area of the Indian state of Gujarat in 2001. Amazingly, the infrared signals were there a clear week or two before the quakes happened. There is much work to be done, however, before both Freund's and

Bryant's work join the mainstream, and controversy still revolves about some of their ideas. Most importantly, questions remain to be answered about whether or not the infrared emissions are wholly reliable. In other words, do they occur before every earthquake? Can they happen without an earthquake following? If the answer to the first question is yes, and to the second no, then the quest for the Grail could be approaching its end with the development of an effective earthquake prediction method capable of monitoring all the world's earthquake zones from space—an incredibly powerful disaster mitigation tool that could save millions of lives, including tens of thousands in the Japanese capital.

Others too claim to be within touching distance of the Grail, this time through examining the pattern of tremors that leads up to a major quake. Geophysicist Vladimir Keilis-Borok and his team at the University of California, Los Angeles, posted a warning in April 2004 that southern California would be shaken by an earthquake of Richter magnitude 6.4 or greater by 5 September of the same year. The team claims to have successfully predicted the magnitude 6.5 San Simeon quake that struck central California just before Christmas in 2003, and the prediction has been endorsed by the California Earthquake Prediction Evaluation Council. While Keilis-Borok bases his prediction on consideration of 'chains' of earthquakes that occurred during preceding years in the region of the main quake, there is as yet no accepted physical explanation for the link. Given the wide temporal and spatial leeway in the prediction—sometime in a five-month period and somewhere within an area of 31,000 square kilometres—the prediction is really more of a fore-cast. It is a broad expression of probability and lacks the required capability to predict an earthquake of a particular size on a particular fault to within a few days. Furthermore, the area for which the prediction has been made is so seis-mically active—two earthquakes of magnitude 7 or more have occurred here since 1992—that there is a 10 per cent chance of a magnitude 6.4 earthquake occurring randomly before the 5 September deadline. Unfortunately, no quake had made an appearance by March 2005, so it seems it is back to the drawing board for Vlad and his team.

Bearing in mind that the battering of the Greater Tokyo Metropolitan Region by a repeat of the Great Kanto quake may be several decades away, prospects look increasingly bright for the timing of the event to be pinned down to within a time-frame of months, weeks, or even days. This should help to reduce the predicted 60,000 plus death toll to just a handful, and limit damage through providing time to make safe the most serious potential sources of fire in advance of the quake. Tokyo Gas is already on its way to doing this through its SUPREME (SUPer-dense REal-time Monitoring of Earthquakes) system, which automatically shuts off the gas supply to districts that have sustained signifi-cant earthquake damage. The scale of destruction can be further reduced by having in place automated early warning systems (EWSs) that provide an alert after the 'big one' has happened but before its most destructive waves reach the capital. All earthquakes generate a range of different shock waves. *P (primary or push)* waves travel the fastest and are the first to be picked up by a seismograph, but they cause little damage. *S (secondary or shear)* waves, on the other hand, are far more destructive but travel more slowly than the P waves and therefore fall progressively further behind as the shock waves move away from the earthquake source. Surface waves are the slowest of all, as they follow a longer path around the surface of the Earth rather than through it. By having seismographs broadcast an electronic warning the minute they detect P wave signals large enough to indicate that a big quake has happened, measures can be taken to substantially reduce its impact. In Japan, such a system—known as UeEDAS—already sends warnings to bullet trains in the event of a significant tremor, causing them to slow down, and similar systems are planned in Istanbul and southern California. Even a warning of as little as 15 to 20 seconds can be sufficient to change traffic lights to prevent traffic entering motorways, allow air traffic controllers time to turn away inbound aircraft, shut down factory production lines, turn off gas pipelines, and save valuable computer data.

Neither accurate predictions nor sophisticated early warning systems will be able to save the Japanese capital unless the majority of buildings are constructed to withstand a quake of magnitude 8 or more. There is a saying among earthquake engineers: 'earthquakes don't kill people, buildings kill people.' At no time was this truism better demonstrated than in December 2003, when two earthquakes of about the same size—in central California and Bam (Iran)—had tragically different outcomes. At San Simeon in California a magnitude 6.5 quake that hit on 22 December damaged just 40 properties and killed two people. Four days later a similar-sized quake destroyed 85 per cent of the buildings in the Iranian city of Bam and took more than 26,000 lives. The astonishing discrepancy between the impacts of the two events can be put down almost entirely to building design and construction. In Iran, most properties were cobbled together from mud brick (adobe), a building material that performs worse during earthquakes than almost any other. In stark contrast, some of the most stringent anti-seismic construction codes on the planet ensured that most properties in California were barely scratched.

The 33 million inhabitants of the Greater Tokyo Metropolitan region live and work in a range of properties, some very old, others very new, some designed to cope with the greatest earthquake, others wobbly enough to collapse in the slightest tremor. The effect of a large earthquake on such a wide range of building stock was vividly demonstrated in 1995 at the city of Kobe, 400 km south of the capital, when a magnitude 7.2 quake damaged or destroyed more than 140,000 buildings. How a building behaved during the event depended almost entirely on its age, with those put up later having to adhere to progressively stricter codes of design and construction. Consequently, all other factors being equal, more older properties were damaged and destroyed than newer ones. The next major Tokyo quake is liable to see a similar pattern. If it can be predicted, however, and that is still a big if, the resulting death toll can be expected to be far smaller than the 60,000 estimate that currently stands. Even without accurate prediction, if the full potential of early warning systems is realized by the time the quake strikes, shutting down gas supplies and making safe other potential sources of fire, then the great conflagrations of 1923 will be avoided. Whether predicted or not, the millions of older, less well-constructed properties will ensure that the scale of damage, following a repeat of the Great Kanto quake, will be immense. Nevertheless, whether the event will have global economic ramifications is likely to depend upon how the commercial centres survive, and this in turn will depend critically upon how well they are put together. The Japanese have led the way in designing and erecting some of the world's most sophisticated anti-seismic, high-rise buildings. Like all good quake-proof structures, they are designed to be strong but flexible so that they can to some extent 'go' with the shaking to minimize strain and therefore damage. In order to accomplish this, some buildings are set on enormous springs or rubber 'dampers' that absorb much of the severest shaking before it gets into the building itself—a measure known as *base isolation*. Others have computer-controlled *dampers* within the buildings themselves, which absorb seismic energy and reduce swaying. Columns and beams are designed to deform in response to shaking rather than breaking apart, while external walls braced by extra cross-beams and additional internal walls provide added strength. If the 'big one' takes several decades to arrive, then new materials and technologies are likely to have been introduced to make newly constructed buildings even safer during major earthquakes. Within decades it would not be surprising, for example, for buildings constructed in hazardous regions to benefit from impregnation with carbon nano-tubes, 30 to 40 times stronger than steel, making them virtually indestructible.

Whenever I give a talk about natural catastrophes, someone in the audience inevitably wants to know if it is possible physically to prevent an eruption, a tsunami, or an earthquake happening. I have already addressed this issue with regard to the first two phenomena, and prospects for doing something about them appear bleak, unless you have enormous faith in future technologies. But what of earthquakes? Can we stop them or at least make them smaller? Well, we already know that human activities have accidentally triggered earthquakes, and examples abound. After the Hoover Dam was constructed in Nevada in 1935, blocking the Colorado River to form Lake Mead, the weight of the water triggered over 600 quakes, some reaching magnitude 5. Similarly, in 1967 a magnitude 6.5 quake triggered by the filling of a reservoir behind India's Koyna Dam, killed up to 200 people. Nuclear tests have also triggered earthquakes of up to magnitude 5, while in the late 1960s, toxic waste pumped into 4 km deep wells at Rocky Flats near Denver, Colorado, initiated earthquakes in an area that had previously been seismically quiet. For some time, scientists have pondered the idea of deliberately triggering earthquakes by lubricating faults—by pumping water into them, in order to persuade them to move before they have accumulated too much strain. So far, however, and perhaps not surprisingly, no one has actually tried it. In the hugely litigious US, it is doubtful if this will now ever be attempted, and as far as I know there is little enthusiasm elsewhere in the world's seismic hot spots. The problem is one of control; how to cajole a fault to rupture to give a quake of just the size required to relieve stress but not to cause damage. For example, if sufficient strain had accumulated along a fault to trigger a magnitude 8 quake, then a lubrication programme would need to trigger over three-quarters of a million smaller, magnitude 4 earthquakes to release all the energy in the fault, a prospect that is clearly not feasible. This could also be done by triggering just 30 or so magnitude 7 quakes, but these could actually cause more destruction, and certainly more disruption, than a single magnitude 8 event. So it looks as though it is back to square one for the quake-starters.

Forgetting the risky business of fault lubrication, the future for the Japanese capital appears perhaps a little less bleak than suggested in the opening chapter of this book. With the prospect of effective prediction offering the possibility of a dramatically reduced death toll, improved early-warning systems helping to minimize the impact of the coming quake on lifelines, and appropriately engineered buildings helping to limit the scale of damage, maybe the economic cost of the 'big one' when it arrives will be far below the estimated 4.3 trillion US dollars. It might even be the case that we are close to crossing this particular event off the terrestrial geegee list. That still leaves us to face the volcanic super-blasts and towering ocean-wide tsunami, so we shouldn't get too blasé. I hope I have been able to convince you, in the course of this chapter, that we do have the capabilities to cope with these seemingly unstoppable events, and if not prevent them, then at least limit their detrimental effects on our civilization. In order to mitigate and manage the worst our dynamic Earth can throw at us, the international community must first find some way of working together for the common good. As I touched on in Chapter 1 and will return to in the next chapter, with respect to climate change this is something we have so far failed spectacularly to accomplish, even though here is a gee-gee that is happening now. It could therefore be quite a while before nations get their act together sufficiently to even begin to think about catastrophic geological events that may not become a problem for centuries or millennia. When this eventually happens, of course, it could well be too late. We need to be tackling the tectonic threat now, following the lead of those addressing the menace from space by systematically identifying and cataloguing potential problem areas, including unstable island volcanoes and continental margins and possible super-eruption sources, and by improving our surveillance of specific, more immediate, threats such as La Palma's Cumbre Vieja. We should also be building computer models capable of accurately replicating the climatic conditions we can expect to face after a super-eruption, or the pattern and scale of tsunami generation following a giant landslide into the ocean. At a national and international level, the issues of raising awareness of the tectonic threat and improving multinational lines of risk communication need to be addressed, along with the establishment of better warning systems and protocols, and the drawing up of contingency plans for relief and recovery. We have the tools to prepare *now*, should we wish to use them. After the events of Boxing Day, 2004 failure to do so would be both criminal and stupid.

4

Walking the Climate Change Tightrope

Climate is an angry beast and we are poking at it with sticks.
Wallace Broeker, oceanographer and climate scientist.

In the summer of 2001, just a few days after hunting for rogue asteroids at Kitt Peak Observatory, I found myself at the world-famous Monterey Bay Aquarium Research Institute on the Pacific coast of California, this time with a view to looking down into the deep ocean rather than up at the stars. I was there to interview a team of oceanographers for my BBC Radio series, *Scientists Under Pressure*. I must say that Peter Brewer and his colleagues did not look particularly pressured as they pottered around a sunny dry dock in their shorts, preparing a complicated looking ROV (Remotely Operated Vehicle) called *Raptor*—a sort of tethered, unmanned submarine—for loading onto their research ship, the *Western Flyer*. There is little doubt, however, that time is running out for Brewer and his team, for the experiments they were about to undertake on the deep ocean floor might offer a solution to global warming, *if* it can be shown to work and *if* it can be taken up on an industrial scale: two very big 'ifs'.

Brewer and his colleagues are just one of many research teams looking into a concept known as *carbon sequestration,* in other words, removing carbon dioxide from the atmosphere and stashing it away where it can't do any harm. The ultimate purpose is to begin to reduce carbon dioxide concentrations in the atmosphere before it is too late. Using the ROV, the Brewer team were about to transport 45 litres of liquid carbon dioxide down to the floor of the Pacific Ocean, where they hoped it would stay for a very long time. At 3 km depth, the liquid carbon dioxide is slightly more dense than sea-water, and so will bob happily along the seabed. Below this depth, it will combine with water to form a solid *gas hydrate* deposit. Hiding carbon dioxide down in the deep ocean might seem like a great idea to rid the atmosphere of the ever-increasing surplus that is accelerating global warming, but only if it can be guaranteed to *stay* down there. According to Brewer, the carbon dioxide does slowly dissolve into the seawater, but when it has done so it becomes a component of the ocean, where it is likely to stay for centuries or millennia. Certainly Brewer is confident that it will not return to the atmosphere for a very long time, if ever. But what about the effect on marine life? As you might expect, this is a big concern at the Monterey Aquarium Research Institute. Releasing carbon dioxide into the ocean has the effect of lowering the pH and making it more acid, a change that might affect some of the estimated 10 million species that live in the ocean depths. Brewer's biologist—Michael Tamburri—has been looking into this by releasing small blobs of liquid carbon dioxide onto the sea floor to find out how it affects the fish, eyeless shrimps, worms, and molluscs that eke out their exist-ence in the deep ocean. Although agreeing that further research is needed, according to Tamburri the effects seem minimal. Fish don't avoid the carbon dioxide-rich areas and will even swim into them if attracted by food. Agreed, they do fall unconscious and float around as if dead, but they soon recover once they have drifted away to a safe distance. So that's all right then.

One of the biggest problems with the Monterey team's solution to carbon dioxide build-up in the atmosphere lies in the enormous technical issues involved in pumping huge volumes of the gas down to a depth of 3 km or more. Else-where in the world, it may be possible to get around this by letting the ocean do much of the work. Helga Drange, a Norwegian oceanographer at Bergen's Nansen Environ-mental and Remote Sensing Centre, is proposing that liquid carbon dioxide would only need to be pumped to a depth of 800 m off the Norwegian coast, whence the dense, cold waters of the region would quickly drag it down to the dark depths of the North Atlantic.

The ocean is the planet's biggest *carbon dioxide sink*, and it sucks in around 25 million tonnes of the gas from the atmos-phere every day. Without this, a runaway greenhouse effect would long ago have taken hold, leading to a barren world simmering happily at a temperature of several hundred degrees C. A little more, you might think, could not possibly do any harm—surely. This may well be true, but there remain immense problems to be solved with regard to ocean seques-tration. There is a widespread feeling in environmental circles that such a techno-fix is simply an excuse dreamt up by the industrial lobby to allow them to go on pumping more and more greenhouse gases into the atmosphere. In addition to concerns over how long the stuff will stay down there—and one suggestion is only 300 years at the most—and the impact on marine life, there are also serious worries over the effect on the environment of the enormous scale of the pumping required for ocean sequestration to have any significant impact on atmospheric concentrations of the gas, which would see pipelines snaking their way seawards from countless industrial and power plants clustered along the world's coastlines. Most of all, a sceptical public will need to be convinced that having seriously damaged the atmosphere, we now need to mess about with the oceans to sort the problem out. In the end, it will almost certainly be this attitude, if anything, which kills off any large-scale plans to dump unwanted carbon dioxide in the open ocean.

There are, however, other places where we can lock it away, and a number of countries, including the UK, the US, and Australia, are looking to store their greenhouse gas excesses in rocks deep beneath the seabed. Unlike ocean sequestration, which allows the carbon dioxide to float about at will, *underground sequestration* involves the gas in liquid form being pumped into sedimentary rocks, where it becomes trapped in the pore spaces between grains in much the same way as oil and gas are contained within a hydrocarbon reservoir. The carbon dioxide is thus completely isolated from the atmosphere. In the North Sea plans are already far advanced, and the Norwegian oil company, Statoil, has been pumping unwanted carbon dioxide gas from its Sleipner field into a porous sandstone layer a kilometre beneath the sea floor since 1996. The amount of space available in the storage rock is enormous, and if just 1 per cent were utilized it could take three years' worth of carbon dioxide output from all of Europe's power stations. The success of the Sleipner project has enthused the UK government, which is proposing to launch a *Carbon Dioxide Capture and Storage* (CCS) scheme that makes use of North Sea aquifers or disused oil and gas reservoirs to help the country reach its target of a 60 per cent cut in greenhouse gas emissions by 2050. It has even been proposed that excess carbon dioxide could be pumped into oil reservoirs that are on their last legs so that the gas forces out the final dregs of the oil. This is an established technique known as *enhanced recovery*, which the oil industry has used for some time. A scheme backed by US recovery company Kinder Morgan already has plans to dispose of around 1.6 billion tonnes of carbon dioxide in a North Sea field in order to squeeze out an extra 5.3 billion barrels of oil. The fact that combustion of each and every barrel releases 487 kg of carbon dioxide to the atmosphere, however, makes a very persuasive argument for leaving the oil in the ground. In any case, the cost of using carbon dioxide produced by industry for reaping the last of the North Sea oil bonanza is prohibitively expensive, with the cost of transporting and using the carbon dioxide from a single power station estimated at over 2.5 billion US dollars.

While the sequestered gas is likely to be more secure in its subterranean bunkers than if left to its own devices on the ocean floor, many obstacles need to be overcome before underground sequestration can be developed on a scale large enough to make a dent in projected emissions. Although any safety problems can probably be overcome, proponents of the solution will need a concerted hearts and minds campaign to win over the environmental lobby and public opinion. Greenpeace, Friends of the Earth, and other major campaigners argue reasonably against underground sequestration on two fronts. First, like marine sequestration, underground storage will require networks of pumping stations and pipelines that will have a detrimental impact on the local environment. Second, and more important, they view any sequestration method as a ruse devised to dodge responsibility to actively cut back on the production of GHGs by adopting a more sustainable outlook. New and untried technologies always come with attached risk, and the potential to make things even worse. In this case, the main concern is that— perhaps decades or centuries down the line—the reservoirs will begin to leak, transforming them from carbon dioxide sinks into sources and thereby contributing to warming rather than helping to reduce it.

There is also a more natural way of sequestering carbon dioxide, which is to plant forests. Here too, however, there are problems, not the least of which is keeping track of the fate of the carbon locked away in new vegetation, much of which can find its way back into the atmosphere due to forest fires. Young trees are also especially susceptible to wind damage, a phenomenon we may well experience far more often as global warming promotes a more erratic climate. The US, in particular, seems keen on this *agricultural sequestration*, some say, because it allows them to balance this against the ever increasing amounts of carbon dioxide they are pumping into the atmosphere. With US cropland soils already having lost between 33 and 60 per cent of their stored carbon since the earliest days of cultivation, proposals have been put forward to replenish this diminished carbon pool through tree-planting, conservation tillage, and better land-management practices. Opponents argue that such a scheme would be largely cosmetic and would hardly make a dent in the enormous contribution of the US to accelerating global warming. Agricultural soils in the US capture between 0.1 and 1 per cent of the country's total annual GHG emissions. Furthermore, agricultural emissions are 25 times higher than the amount that is locked away in the soils. In any case, replenishing soil carbon could only be a short-term fix, with the soils likely to become saturated within decades, thereby becoming a source for carbon dioxide rather than a sink. And there is a sting in the tail too. Planting trees has always been regarded as a good thing, particularly since global warming came along. They certainly play an enormous role in extracting carbon dioxide from the atmosphere and locking it away in their woody frames, but there is a twist to the story. Trees are also very effective at stabilizing the surface, tying down soils, and transforming dust bowls to lush meadows—so more trees, less dust. Surely that's a good thing, you say? Well, dust from the continents is carried out over the oceans where it settles, providing microscopic marine algae—known as *phytoplankton*—with the supply of iron they need to grow and reproduce. Less dust—less phytoplankton—and *more* carbon dioxide in the atmosphere. This is because these tiny organisms are great devourers of the gas, which they use to build their skeletons. Research undertaken by Andy Ridgwell, an environmental scientist formerly at the University of East Anglia, and now at the University of California, Riverside, suggests that just a 20 per cent fall in the supply of dust reaching the oceans could reduce the uptake of carbon dioxide by phytoplankton by 150 million tonnes a year—just about equivalent to the annual production of the gas by the UK. Furthermore, it seems likely that the loss would more than outweigh the amount of carbon dioxide absorbed by the new trees that caused the dust reduction in the first place. With China planning a massive reforestation programme on a scale that only it can manage, worries are already being expressed that planting forests may actually accelerate warming rather than helping to apply the brakes.

If we want to keep any grip at all on the rate at which the planet is warming up, then it is crucial that we keep the phytoplankton happy and well fed, because these innocuous, microscopic creatures are responsible for close to half of all the photosynthesis on the planet—the process whereby plants use energy from sunlight, water, and carbon dioxide from the atmosphere to make their food. One way of keeping them not just happy but ecstatic is to provide those living in parts of the sea where the iron they feed on is scarce with additional food supplies. The reason for doing this does not reflect any particular concern for the plankton's welfare, but is a way of trying to get them to increase their uptake of carbon dioxide from the atmosphere. Providing more food should encourage them to reproduce rapidly, resulting in an algal 'bloom' that will suck in more of the gas. When the organisms eventually die, the idea is that they will sink to the sea floor, thereby isolating the carbon from the atmosphere. Unfortunately, small-scale experiments involving seeding of parts of the southern oceans using iron sulphate have proved inconclusive, and there is much scepticism about how such seeding can be undertaken at a scale big enough to make a worthwhile difference to carbon dioxide concentrations in the atmosphere. Parts of the ocean would need to be seeded with iron on an industrial scale *every year*, and even then there is no guarantee that a reduction in atmospheric carbon dioxide would result. Instead of the creatures dying and sinking to the seabed, they could be eaten by larger marine invertebrates, which in turn could be swallowed up by fish and other sea creatures who would release the gas back into the atmosphere during respiration. In this way, instead of lowering carbon dioxide concentrations in the atmosphere, seeding the oceans with iron might actually have the reverse effect.

It seems, then, that the whole business of carbon dioxide sequestration—a simple concept at heart—is proving to be a highly complex issue beset with problems and potential pitfalls. Nevertheless, with Kyoto struggling and GHG emissions still on a steeply rising curve, it is an approach that we may well be forced to embrace in one form or another, not to take the place of, but to complement, reforms to slash the production of carbon dioxide and other GHGs. Whatever you think of their true motives, most advocates of sequestration at the very least acknowledge that global warming represents *the* great challenge of our age, and are coming up with what they see as workable solutions.

Sequestration is just one of a number of ways in which we can attempt to bring global warming under control. Each, on its own, is unlikely to have the desired effect, but together they certainly have the potential to mitigate and manage the problem and eventually bring it to heel. The key to slowing the rate of warming is a significant reduction in GHG emissions, which—despite Kyoto—are still very much on the up. Emission cuts on the scale needed to make any noticeable dent in the concentration of GHGs in the atmosphere are going to require dramatic changes in the way those of us privileged enough to live in industrialized countries lead our lives, and sacrifices will have to be made if we are to reap the benefits of a return to a temperate world. The developing nations are also going to have to play their part, curbing—as much as they are able—the headlong gallop for unsustainable growth that has galvanized the economies of the developed countries since the Industrial Revolution began to take hold in seventeenth-century England. The move away from fossil fuels and towards alternative, renewable energy sources has a critical role to play in bringing down the level of GHG emissions, and later I will look at the varied and growing list of potential power sources that range from wind to tap water. With the latest forecasts suggesting that global temperatures may be on course to rise by 7–10 degrees C by the end of the century, and the effects of warming already beginning to hit home, we also have no option but to adapt to a warmer planet. In the industrialized countries this will involve lifestyle changes and compromises that may be little more than inconvenient. In the developing world, however, they are likely to take the form of the wholesale movement of parts of the population and drastic changes in agriculture.

To *reduce* the atmospheric concentration of GHGs and *adapt* to a much warmer world seems then the only reasonable way we can reverse the warming trend. There are, however, less reasonable points of view. The first—and incredibly it still exists in some circles—can be summarized by the question 'What global warming?' Yes, despite the huge body of evidence to the contrary, a tiny minority of ostriches still holds either that global warming is not happening or that it is nothing whatsoever to do with the fact that human activities have increased GHG concentrations by more than a third over the past few centuries. The second view is more gung-ho than head-in-the-sand, and is held by those ever-optimistic techies who are forever trying to outdo one another increasingly more outlandish schemes for defeating global warming at a stroke. Taking a lead from a recent *New Scientist* editorial, I will simply refer to them collectively as the 'are you feeling lucky' school, but there will be more of their weird and wonderful ideas presently.

To return to the ostriches, I have always been fascinated by their thought—or should that be lack of thought—processes and by the way they rationalize their refutation of a phenomenon that is undeniable. Undoubtedly, some who belong to this school have a purely Machiavellian agenda that will accrue benefits should the world continue to warm; others, however, present arguments to support their case that are so ludicrous or convoluted that you worry for their proponents' sanity. A particularly good one—and a favourite of some elements of the tabloid press—is that the whole global warming idea is a conspiracy dreamt up by climate scientists to increase funding for their research. In a diatribe along these lines, which displays astounding ignorance of the science and a complete and utter misunderstanding of the issues, one well-known commentator in a right-wing UK tabloid has denounced global warming as a 'global fraud' foisted upon an unsuspecting world by a 'left-wing, anti-American, anti-west ideology'. Another viewpoint holds that humans are far too inconsequential for our activities ever to have a significant impact on the climate, *ergo* global warming must be a natural effect. It would be enough to make you weep if the situation was not so desperate. The truth is that human activities have an enormous influence on the climate. In fact, in a recent paper, climate scientist William Ruddiman, of the University of Virginia at Charlottesville, argued convincingly for our ancestors instigating the warming trend as long as 8,000 years ago, when—following the end of the last Ice Age—GHG concentrations in the atmosphere started to rise more rapidly than expected. Ruddiman suggests that this might have coincided with the birth of agriculture, with forest clearances in Europe, India, and China raising levels of carbon dioxide, and rice paddies and burgeoning livestock numbers pumping out increasing amounts of methane. It seems then that we are now eight millennia into what some scientists are calling the *Anthropocene*, perhaps soon to be recognized as the latest epoch in a geological timescale that stretches back 4.6 billion years, and the only one to be utterly dominated by a single species with the will and the capability to mould the planet. If *we* make a mistake, the entire planet and everything on it suffers—and we have made a big one.

Let's return then to consideration of how we can get things back on track before it's too late. Carbon sequestration shows some promise as a short-term fix, but on its own it can't halt the rise in temperatures. The key, as I mentioned earlier, is to reduce GHG emissions as soon as possible and by as much as we can. With the biggest polluters holding back, the Kyoto goal of cutting emissions by 5.2 per cent below 1990 levels by 2008–2012 is never going to be achieved. So long as the United States emits 25 times as much carbon dioxide per head as India, and a Bangladeshi accounts for one-fiftieth the emissions of a Brit, there is little hope of progress along the lines proposed in the Kyoto Protocol. The problem is that countries like the US, Canada, and Australia are refusing to take any action that may hinder their economies, while allowing developing countries such as China and India a free hand to increase GHG emissions from their comparably low levels. The polluters argue that if they have to cut back then so should everyone. Understandably, the developing countries are none too keen, viewing this stance as a conspiracy to keep them and their economies from expanding.

Even with Kyoto seemingly too little too late, there is still hope that an international agreement for stabilizing and ultimately reducing GHGs can be arrived at, with increasing attention being focused on a new way forward known as *Contraction and Convergence* or simply C&C. Developed by the London-based Global Commons Institute, C&C seeks to share out the burden of cutting GHG emissions fairly across all the nations of the world by reducing emissions on a per capita basis. In other words, following an international accord on an agreed level for all emissions—perhaps reviewed every five years or so—the amount each country will be allowed to emit will depend upon the size of its population. If a country is unable to utilize its entitlement, then it is free to trade this with another state that might wish to, or have to, increase its emissions beyond its allocation. At the turn of the millennium, the world's population stood at about 6 billion, the same figure (in tonnes) as the total amount of carbon dioxide in the atmosphere. If C&C had been in operation then, everyone on the planet would have had the right to emit a tonne of carbon dioxide each. The reality was far different, however, with the US pumping out over 5 tonnes of the gas for every citizen, compared with, for example, 0.09 tonnes for every Nigerian. Ultimately, the goal is to reduce the average annual output of every man, woman, and child on the planet to around a third of a tonne. The Global Commons Institute is currently proposing that when and if adopted by the international community, C&C should—in the first instance—work towards stabilizing the carbon dioxide concentration in the atmosphere at 450 ppm by the end of the century. This is around 80 ppm higher than current levels but far lower than a 'business as usual' scenario, which would see concentrations topping 800 ppm by 2100. This *contraction* in emissions is critical if the worst ravages of global warming are to be avoided, but even these may not be enough. With increasing signs that global warming is already beginning to bite through increased climate variability, it may be that this target will have to be revised downwards if the situation deteriorates dramatically. The *convergence* bit, whereby entitlement to emit GHGs is equalized entirely on a per capita basis, will—not surprisingly—take a considerable amount of heated bargaining to sort out, but the Global Commons Institute is presenting 2045, the year of the UN Centenary, as a possible target for adoption.

According to the 2001 forecast of the Intergovernmental Panel on Climate Change, a carbon dioxide concentration in the atmosphere of 450 ppm in 2100 should result in a global average temperature rise of 1.4 degrees C. This is more than twice the rise of 0.6 degrees C that occurred during the last century, but not too high to make effective adaptation impossible. But although C&C is attracting considerable and growing support, it is still a very long way from being adopted by the international community. Bearing in mind as well that temperature rises during the coming century may in fact be far higher than expected, we certainly can't afford to rest on our laurels.

Pretty much everyone in the know is agreed that GHG emissions need to be cut dramatically, but how to do it? Clearly, national governments and multinational bodies like the European Union need to take the lead in launching policies that bring this about, but there is also a great deal we can do as individuals. On the personal level, we can make a difference, for example, through choosing to holiday at home rather than jetting off across the planet, or by eating locally grown produce. At the national or multinational level, the focus must be on the development of renewable power sources at the expense of fossil fuels, and on increasing energy efficiency.

The UK is committed to reducing the country's GHG emissions by 12.5 per cent on 1990 levels and is already close to this mark. Carbon dioxide emissions are 9 per cent lower than in 1990, despite economic growth close to 30 per cent over the same period. This is a clear indicator that the argument constantly put forward by the Bush administration and others—that the economic cost of coping with global warming is simply too much to bear—is a fallacy and a smokescreen. In January 2000, the UK became the first major industrial economy to announce caps on GHG emissions by heavy industry such as oil and gas refineries, power stations, and steel works, with an average target reduction of 16 per cent by 2010. The move is designed not only to further curb emissions, but also to provide incentives for energy companies who turn away from coal, oil, and gas and embrace cleaner, renewable energy sources. Under a new EU GHG emissions trading scheme, such companies will be able to sell 'permits to pollute' to other industrial concerns that are unwilling or unable to meet their targets. Inevitably there are detractors, and a segment of UK industry is bemoaning the supposedly terrible effect the measures will have, forecasting draconian rises in electricity prices of up to 80 per cent by the end of the decade, by which time the UK has pledged to reduce its GHG emissions to 20 per cent below 1990 levels. As a 12 per cent fall has already been accomplished without noticeable pain, however, this is surely plain, unadulterated scaremongering by the fat cats who don't want to see the value of their bonuses slip.

The government chief scientist, David King, has indicated that the UK is committed to cutting emissions by a further 10 per cent every decade until 2050, but how might these additional reductions be accomplished? The key has to be increased investment in renewable power sources. For an island nation as blustery as ours, it is a crime that until recently our love affair with wave and wind energy has been such a capricious one. Now, however, we are forced to take the plunge, if for no other reason than to ensure security of energy supplies in coming decades. In barely 15 years' time, the UK could be dependent on imported energy for three-quarters of its primary needs, much of it natural gas transported by exposed pipeline across thousands of kilometres from politically unstable Central Asia. Despite recent murmurings about a revival, the nuclear option cannot plug the gap. It is too expensive, new power stations will take too long to construct, the waste issue has still not been resolved, and public opinion wouldn't stand for it. In an attempt to remove at least some eggs from this highly vulnerable basket, and to further cut back GHG emissions, the government announced in 2003 its goal of achieving 10 per cent of power from renewable sources such as wind and waves by 2010, and 20 per cent by a decade later.

12 The UK is the windiest country in Europe, so wind is an energy resource we cannot afford to ignore. A miniature wind turbine developed by a Scottish company, Windsave, can provide up to 33 per cent of annual electricity costs for an average house and costs just less than £1000.

While they are greeted with scepticism in some quarters, there is no question that alternative power-generating schemes are taking off in a big way, with wind power at the forefront. As the windiest country in Europe this was inevitable, but there is still a battle to be fought before wind can be counted a major supplier of UK energy. Onshore wind farms have been branded noisy and unsightly and are highly unpopular with the Ministry of Defence as they can interfere with radar. Offshore schemes have raised the ire of fishermen and the Royal Society for the Protection of Birds has expressed concerns about the potential threat of the flailing vanes to flocks of migrating birds. Offshore wind farms do seem to be the way forward, however, and the UK has recently joined the exclusive club of countries operating large-scale offshore wind farms, which includes Denmark, Sweden, and Ireland. It is just 13 years since the first offshore farm was built at Vindeby in southern Denmark. Since then, however, wind power has gone from strength to strength. In Denmark itself, wind now supplies 20 per cent of all energy and the figure will rise to 25 per cent in just five years. Europe in general is leading the way in offshore wind power. Plans are afoot for floating wind generators able to operate in waters as deep as 200 m, and with wind-generated electricity forecast to become—by 2010—as cheap as that produced by gas, it looks as if the future of the industry is secure, not only in wind-battered north-western Europe, but anywhere on the planet where the wind is sufficiently reliable.

As well as being the windiest country in Europe, the UK

can also boast the biggest tides in the world and some of the stormiest seas. Once again, however, these almost limitless sources of clean, renewable power have been left to sit on the sidelines while coal, oil, gas, and nuclear power have kept the game to themselves. Today we are at the same stage we were in the early 1980s, when wave power was all the rage and it seemed only a matter of time before our kettles and computers were powered by the friendliest of environmental energy from the rocky coasts of Cornwall or the islands of north-west Scotland. It was not to be, however, and for a very silly reason. In a government report evaluating how much electricity from wave energy would cost, someone—for malicious reasons or through plain carelessness—moved a decimal point one place, making it ten times more expensive and therefore not worth bothering with. Two decades later and wave power is back and competitively priced at an estimated 4p a unit—just a third more than coal- or gas-generated electricity, but without the associated pollution. Plenty of problems remain to be solved before the power of the sea can start to make a noticeable contribution to the UK's energy needs, not least how to build a structure rugged enough to survive the constant battering of the North Atlantic. This has already taken its toll, and in 1995 an innovative 1 megawatt wave-power scheme off the Scottish coast was reduced to junk even before it was completed. Severe storms also conspired to postpone by two years the opening of the EU's experimental wave-energy scheme in the Azores. With the sea covering two-thirds of our planet, it is hardly surprising that the World Energy Council has estimated that waves have the potential to supply 10 per cent of the world's energy needs. So far, however, only a handful of wave-power generators are operating successfully. In the UK just a single prototype generator on Islay in the Inner Hebrides is supplying energy to the national grid, although there are now ambitious plans afoot to connect up many more. A test site in the savage seas close to Stromness in the Orkneys has just opened, which will allow up to four different types of wave generator to compete head to head as they pump electricity into the grid. Meanwhile, at the other end of the country, a so-called *wave hub* is to be built 15 km off the south-west coast, providing an array of giant submarine sockets into which manufacturers can plug their wave generators, testing them while at the same time supplying electricity to Devon and Cornwall. If successful, the power of the sea could be supplying the region with up to 7 per cent of its energy needs by 2020. Hot on the heels of wave power's rebirth are plans to tap the power of the enormous tidal ranges that characterize the UK's estuaries and offshore islands. Having been talked about for a quarter of a century, a power-generating barrage across the Severn estuary is once again under consideration, with the huge 13 m tidal range having the potential to supply the UK with 12 per cent of its electricity.

While wind and water are likely to bear the brunt of future power production as the UK progressively reduces its reliance on fossil fuels and brings down its GHG emissions, warmer climes may well turn to the Sun to meet their twenty-first-century energy needs. Solar power remains relatively expensive, but costs are dropping fast and the potential is huge. The oil giant, Shell, has calculated that renewable energy has the potential to supply with ease a world of 10 billion people, with solar power dominating a mix of different renewable technologies. Not to be outdone, rival BP has worked out in a separate study that if all suitable roofs were surfaced with solar *photovoltaic* panels, which convert sunlight to electricity directly, then the Sun could supply more than the current energy needs of the gloomy and cloud-bound UK. In sunnier lands, efficiency would be considerably greater, offering the potential to power not only the villas of the rich in Monte Carlo or Palm Beach but also the homes of billions of impoverished people across the planet. As Jeremy Leggett, chief executive of Solar Century—the UK's largest solar electric solutions company—points out, if made available to the inhabitants of developing countries, solar photovoltaic energy would actually work out cheaper, not only than kerosene but even candles! Other schemes to harness the Sun's energy are also showing their potential. In southern Spain, for example, a thermal solar power station uses 300 huge mirrors to focus the Sun's rays onto a 100-m tower, within which the air is heated to 1,000 degrees C and used to flash water into steam that drives a 1-megawatt turbine. Meanwhile, at Imperial College London, Lund University in Sweden, and elsewhere, teams of biochemists and physicists are going right to the heart of the matter by trying to work out just how plants use sunlight to convert water to energy, with a view to mimicking it. While progress is slow, the reward of success is stunning: the ability to use the Sun's rays to generate limitless amounts of hydrogen—regarded by many as the fuel of the future—both cheaply and without pollution.

Like wind and wave power, solar power has already arrived, with over a third of a million Japanese homes on course to be powered by the Sun by 2005 and 140,000 in Germany. Waiting in the wings, however, are a whole gaggle of new potential energy sources, some humdrum, others exotic in the extreme. Firmly in the former camp come schemes for deriving power from *energy crops*, such as surplus straw, chicken litter, forestry waste, or purpose-grown coppiced trees, from the methane gas generated by landfill sites, and even from human sewage. Heading up the exotic tendency is the dream of a world powered by the aforementioned cheap, clean hydrogen. A highly flammable gas given a bad name in 1937 by tragic images of the blazing wreck of the hydrogen-filled German airship, the *Hindenburg*, hydrogen went through a burst of popularity as *the* fuel of the future just a few years ago. Enthusiasm has cooled somewhat, as more attention has focused on the practicalities of using the gas as a fuel, and doubts about a future global economy driven by clean hydrogen rather than GHG-producing coal, oil, and gas are becoming more widespread. Hydrogen is converted to electricity in *fuel cells*—a British invention that first came to most people's notice when an exploding fuel cell crippled *Apollo 13 en route* to the Moon. Fuel cells have been powering prototype and experimental vehicles in the UK and elsewhere for some time, and in Iceland, which is vowing to abandon fossil fuels and become the world's first hydrogen economy, filling stations to serve hydrogen-powered cars, trucks, and buses have already opened. Within 30 years the country intends to switch from petrol and diesel to hydrogen, making the gas by using renewable hydropower and geothermal energy sources to produce clean electricity that is then used to split water into oxygen and hydrogen. The gas will power fuel cells that not only drive road vehicles but also the country's fishing fleet; and with the only by-product being water, there are no GHG emissions either.

Does this sound too good to be true? It may well be. In the UK and other major industrial economies, most hydrogen—at least for the foreseeable future—would need to be made using electricity generated by burning fossil fuels, so production of the fuel would still be making a contribution to global warming. The gas will also have to become far more cost-competitive before we all rush to the showrooms to demand a spanking new hydrogen-powered SUV or hatchback. Furthermore, potential environmental problems are coming to light that look like taking some of the shine off the gas's bright new image. It seems that this supposedly non-polluting fuel has a penchant for escaping, not only during production but also during transport and when in storage. In fact, it is estimated that around 10 per cent of all hydrogen manufactured would find its way into the atmosphere, where it would begin to damage the ozone layer. In a global hydrogen economy, escaped hydrogen would total 60 million tonnes a year, slashing ozone levels over the poles by up to 8 per cent. All is not lost, however, and a new technique for producing hydrogen from common-or-garden alcohol may do away altogether with the need to transport and store the gas itself, conversion from alcohol to hydrogen taking place at the point of use when it is needed. This revolutionary technique also has the potential to enable developing countries to embrace a hydrogen economy, with ethanol made from fermented surplus crops, such as maize, being converted to hydrogen for about the same cost as producing petrol. Ultimately, despite the potential problems and issues it raises, it is looking increasingly likely that hydrogen will be the successor to petrol, taking on its mantle sooner as it becomes more economic or later as fossil fuel sources become ever more scarce and expensive to extract.

By far the biggest source of hydrogen in the solar system is the Sun—sufficient in fact to fill around a million Earth-sized storage vessels. Here it drives the process of nuclear fusion, whereby the nuclei of hydrogen atoms fuse to form helium plus stupendous amounts of energy. For as long as I can remember, scientists have been trying to replicate and harness the fusion process in order to provide us with unlimited power from the most ubiquitous resource on the planet—water. In the 1960s I remember poring fascinated over a picture-book that enthused about how fusion would transform our world and our lives and would be on the scene in just 30 or 40 years. Well, here we are 40 years on and more, and still no fusion. In fact, it seems that we are actually no closer now than we were in my childhood, and pundits are still talking of the first commercial fusion reactor being half a century away, if not further. The struggle to develop fusion power has been beset with problems, most related to the immense difficulties involved in trying to contain and manipulate, at a temperature of 100 million degrees C what is, in effect, a tiny piece of a star. I am reliably informed by those in the know, however, that solutions have now been found for most of the particularly intractable problems, leading to the imminent construction of ITER, the International Tokamak Experimental Reactor, in either Japan or France (they are still arguing!). If this produces, as expected, ten times more energy than it consumes, then it could lead to the construction of the first commercial fusion reactors in just— you guessed it—30 to 40 years' time.

But forget big-bucks fusion; what about the possibility of building a miniature fusion reactor on your dining-room table for the price of a few nights down the pub? This whole saga started in 1989, when Martin Fleischman of Southampton University and Stanley Pons of the University of Utah astounded the scientific world, by announcing that they had managed to trigger fusion in a jamjar at room temperature. They claimed to have discovered *cold fusion* by inserting a couple of electrodes made of the precious metals platinum and palladium into a jar of heavy water (normal water in which the hydrogen is replaced by its isotope deuterium) containing some lithium salts. Fleischman and Pons reported that they were getting out up to ten times as much energy as heat than they were putting in as electricity, and claimed that, as this could not be the result of a chemical reaction, it must be the result of nuclear fusion. You can imagine the reaction of those worn and weary scientists who had for three or four decades been struggling to bring about a similar reaction in billion-dollar facilities. Nevertheless,

13 A hydrogen-powered bus refuelling station in Reykjavik, Iceland.

dozens of research groups flocked to try and duplicate the results, but with no success. In the years that followed, Fleischman and Pons vanished into obscurity and cold fusion disappeared from the tabloids and even the science magazines. It never really went away, however, and researchers in the field have been beavering away for the last 15 years or so to try and reproduce the Fleischman and Pons results. Now cold fusion is back in the limelight, cast there by the niggling feeling that there might be something in it after all. The US Department of Energy is certainly convinced and is pledging to review all the evidence prior—possibly—to funding a batch of new cold-fusion research projects. One of the experiments that the DoE will undoubtedly be reviewing is that announced in the journal *Physical Review* in spring 2004 by Rusi Taleyarkhan and his team at Oak Ridge National Laboratory in Tennessee. Like Fleischman and Pons a decade and a half earlier, Taleyarkhan and his colleagues caused a storm when they claimed to have triggered cold fusion by blasting a jar of acetone (nail-varnish remover!) with sound waves. Apparently, the sound waves cause tiny bubbles in the acetone to collapse so violently that they force hydrogen nuclei in the liquid to fuse together, generating energy. The jury remains out on this latest foray into the world of benchtop fusion, but wouldn't it be great if it turned out to be true? Cleaner energy and cleaner nails all in one go.

I could go on all day trawling through the various renewable energy sources that have been or are being touted as the successor to fossil fuels and the conqueror of global warming, and I have no doubt that there are many I have left out. There is just one more that I want to touch on, however, mainly because it is something we all, at least in the industrialized countries, have access to—tap water. If Coca-Cola can try to sell it in the form of superior bottled water, then surely it must contain hidden properties. Larry Kostiuk of Canada's University of Alberta certainly thinks so. In fact, his experiments have convinced him that tap water may provide a source of clean, renewable energy to rival the wind and the sea. Like many of the greatest scientific discoveries, Kostiuk and his colleagues came across the power-generating properties of tap water by accident, while investigating what happens when water is squeezed through extremely narrow glass tubes. (Who knows exactly *why* they were doing this?) What happened was that a small electric current was generated which, when thousands of tubes were bundled together, produced enough electricity to power a light bulb. Although I don't have the space to discuss in detail how the current is generated—it is all to do with positively charged water atoms travelling through the tubes—the potential applications are enormous. Kostiuk sees a world powered by batteries containing water rather than today's corrosive or toxic chemicals, even to the extent of hydro-powered mobile phones and laptop computers. Water-pumping stations and filtration plants could double up as power generators, with watercourses of all sizes churning out electricity for local use or for plugging into a grid.

Developing renewable energy sources that are a viable alternative to fossil fuels is certainly a key weapon in the battle to defeat global warming and, as you will now appreciate, there are plenty to choose from. A Darwinian contest over coming decades will see the survival of the fittest, with a mix of the most competitive sources dominating the power-generation scene of the mid to late twenty-first century and the rest falling by the wayside or playing a minor supporting role. In tackling global warming, however, it is not sufficient simply to switch to cleaner, greener energy. As our planet's population heads towards a forecast peak of around 9 billion in 2070 or so, and more and more nations become industrialized, we must also seek to use less of it. There are a number of ways this can be accomplished, most involving increasing energy efficiency and many of which we can take on ourselves. The most obvious entail improved heat retention in our homes by purchasing A-rated appliances and through insulation and double-glazing. Mind you, as the world warms, keeping heat in is going to be less of a problem than keeping it out, and even in the UK growth in the air-conditioning market threatens to cancel out energy efficiency measures. In the US, the situation is far worse with over 50 per cent of all homes and 98 per cent of new cars having air-con. Some American campers even have air-con units in their tents! The UK appears to be following a similar path, with sales of commercial air-con units rising from 100,000 in 1995 to 400,000 by 2001. More blistering summers like 2003 and sales figures will inevitably rise ever more steeply. India, China, and other developing economies also covet air conditioning, with the result that as the world gets warmer we will expend more and more energy on trying to keep cooler.

There are other ways of cutting back on GHG emissions

than just through increased energy efficiency. The most rapid rise in emissions is currently coming from the extraordinary expansion in air travel, with passenger kilometres flown globally predicted to treble in 30 years or so. By 2050 a billion passengers a year are forecast to be passing through UK airports, with aircraft emissions by this time expected to account for three-quarters of the country's GHG production. This doesn't have to happen, however. Taxing aviation fuel—which, unlike fuel for road and rail transport, is currently exempt from a levy—would make flying more expensive and significantly reduce the rise in the number of passengers and flights. By flying just 2,000 m lower than they do at present, planes would burn 6 per cent less fuel, and by avoiding queueing and flying more direct routes, a further 10 per cent saving could be made. The fact that new aircraft look like being 'greener' will also help. In Europe, the civil aviation industry aims to reduce carbon dioxide emissions by up to 50 per cent by 2020, while in the US the new Boeing 787 Dreamliner is being touted as a clean machine that will use 20 per cent less fuel than other similar-sized passenger aircraft. US biochemists are even assessing the potential of soya oil as an aviation fuel. It might make airports smell like fish and chip shops, but it is *carbon neutral*, meaning that burning it in aircraft engines will simply return carbon dioxide to the atmosphere that the soya plants extracted from it a few months or years earlier. Looking further down the line, designs are being bandied about for lighter, less polluting planes with a smaller environmental 'footprint'; *flying wings* that minimize drag and therefore cut back dramatically on fuel; and even planes that 'morph' their wings to maximize fuel efficiency in different conditions and during different phases of a flight.

So, if you want to save the planet, fly less. I was astonished recently at a television holiday programme on eco-tourism, which presented as environmentally friendly a number of holiday packages that involved travelling halfway around the world, without any mention of the quantities of GHGs emitted *en route*. Don't be seduced, every passenger on a long-haul flight produces 124 kg of carbon dioxide for *every* hour of the trip. An aside here—the 65,000 delegates and their many hangers-on who flew to the last Earth Summit in Johannesburg, where climate change and what to do about it topped the agenda—together emitted around 500,000 tonnes of carbon dioxide. It would take a million Indians or 135,000 UK drivers a year to equal this. Was it worth it, I wonder? Instead of jetting off to Thailand then, walk in the Scottish Highlands, 'do' Europe by rail (taking the Eurostar results in an individual generating just a third of the GHG emissions produced by flying to France), or just stay at home and tend the vegetable plot. Far better to grow your own produce than have it flown to your table from the other side of the planet.

There is plenty more we can do as individuals to help the UK progressively cut its emissions, much of which I am certain you are aware of and are, in fact, probably sick of hearing about. Use public transport wherever possible, resist driving a Humvee, lose the patio heater, plant more trees. This last is the key element in the quest of the most environmentally aware to become carbon neutral, striving to make certain that the amount of carbon dioxide they emit during their (usually celebrity) lives is matched by that absorbed by trees they have planted or caused to be planted. In order to 'neutralize' the carbon emissions of their latest album (apparently more than a kilo of carbon dioxide is emitted for every CD manufactured), the British band, Coldplay, have funded the planting of 10,000 mango trees in India. The potential flaw in this plan, if you remember, is that such wholesale planting might reduce the dust supply to the oceans, thus limiting one of the food sources of the phytoplankton, the greatest carbon-dioxide scroungers of all. Furthermore, if reforestation stabilizes the dust and prevents it entering the atmosphere, this could also lead to further warming, as— like volcanic sulphur aerosols—the tiny dust particles are especially effective at blocking solar radiation and reducing temperatures at the surface.

While the UK government's target of a 60 per cent GHG reduction by 2050 is admirable, certainly by the standards of most other states, it is not ambitious enough for some. In particular, Keith Tovey, at the prestigious School of Earth Sciences at the University of East Anglia, charges the government with dithering when it comes to providing information to the public about how they can help accomplish this. In response, he and like-minded colleagues have set up the carbon-reduction project, Cred, through which they pledge to achieve a 60 per cent GHG reduction in the 6.5 million tonnes of carbon dioxide generated by residents of Norfolk and Suffolk—and in half the time. The Cred team organizes evening classes to teach people how to make roof solar panels that provide hot water; they advise on energy conservation and green ways to travel, and undertake environmental audits for local companies. They are even trying to persuade the Anglican Church to plaster the south-facing roofs of their 600 East Anglian churches with solar panels (all churches are aligned east–west leaving a great expanse of sloping roof facing south) and sell the excess power to the grid. Richard Starkey and Kevin Anderson at the Tyndall Centre for Climate Change Research at Manchester's UMIST go even further than Tovey, proposing that carbon-dioxide emissions in the UK be rationed, with each household allocated an amount they can emit, according to the number of householders over 18. Under the scheme, every adult citizen would have a 'carbon card' which they would hand over when purchasing petrol, paying an electricity bill, or even buying a tree for the garden. A national database would add up a household's account and would also permit trading of carbon units, allowing individuals or households to purchase more or sell those that were unused. For example, a family that uses public transport is likely to fall short of its allocation while a household with two SUVs will probably need more than its allocated share. The value of each carbon unit will depend upon the emission target set by the government, which would be reduced every year until the target of a 60 per cent reduction was reached in mid-century. At the moment, there are no plans to implement the proposal, and the required infrastructure probably doesn't exist at present. The population register, which has to be established for the national identity-card scheme, should, however, make the carbon unit plan workable, with your ID card doubling up as a carbon card.

If and when our carbon emissions are rationed, it might be worthwhile investing in a domestic system that produces heat and electricity at home, reducing the need to tap into mains power supplied from coal, oil, or gas power stations. Using a design 190 years old, the UK-designed microchip unit will use a Stirling engine to generate electricity and heat with an efficiency of 85 per cent, compared with the 35 per cent achieved by power stations. Another method, developed by Scottish company, Windsave, uses a miniature wind turbine attached to the roof, which can provide up to 33 per cent of the annual electricity costs of an average house for an outlay of less than £1000. The generator is already proving highly popular, and sales to small businesses have topped 12,000.

Whatever measures or combination of measures we take, it is now virtually certain that the level of GHGs in the atmosphere at the end of the century will be at least double what they were prior to the Industrial Revolution, and possibly much higher. Global average temperatures will continue to rise, by at least 1.4 degrees C, and we will have no option but to adapt the way we live to the new and ever-changing conditions that a warmer world will bring about. In the UK this will encompass everything from the crops we grow, the food we eat, and the clothes we wear to major modifications to our energy, transport, health, and housing policies. Already, record summer temperatures in recent decades have resulted in farmers growing more sunflowers and maize at the expense of wheat and other traditional crops, while the increasing frequency of summer droughts is making irrigation more and more essential, especially in East Anglia. As temperatures rise so our fish of choice—cod, haddock, and plaice—will head hundreds of kilometres further north to be replaced by mullet, black bream, and octopus. Strangely, having to nip down to the local chippie for mullet and chips might be the thing that finally brings home to people just how much global warming is beginning to affect their lives; that and the fact that in future years we may be washing it down with a passable English red. As well as modifications to our diet, we are going to have to come to terms with a shift in some health priorities. Cold-related deaths will fall, to be replaced by a rapid rise in fatalities due to heat stress. The incidence of skin cancer will increase, while malaria may once again be routinely treated in the UK. While high temperatures and drought will become endemic, extreme rainfall events will also result in a huge increase in the flood risk. In its Foresight report, *Future Flooding*, the Department of Trade and Industry has warned that if flood management policies are not modified to take account of climate change, then the annual cost of damage due to flooding could rise 20 times, from the current 1.5 billion pounds to 21 billion pounds by the 2080s. The number of people at risk is expected to rise, over the same period, from 1.6 million to 2.3–3.6 million, requiring spending of between 20 and 80 billion pounds over the period just to maintain the current level of risk. Many new homes constructed in threatened areas will need to have anti-flood measures built in, including airtight covers that fit over airbricks and other external open-ings, and power points set halfway up the wall. Other houses may have to be constructed on stilts for protection from flood waters. Ironically, in other parts of the country, homes will require major renovation in order to counteract damage due to subsidence caused by the drying out of underlying clay soils during the ever more sweltering summers. Worsen-ing of the UK flood problem is not simply a matter of coping with more extreme precipitation causing rivers to overtop their banks; coastal flooding is also predicted to be an increasingly serious issue, particularly in south-east England. Here a combination of continued sinking of the land, which has been going on since the ice retreated from the north 10,000 years ago and set the UK tilting southwards, rising sea levels, increased wave heights, and higher storm surges will make many current defences ineffective by around 2030. The Thames Barrier and the associated 32 km of coastal defences will need to have been increased in height by this time to prevent major flooding during storm surges along the coasts of south Essex and north Kent.

In order to adapt successfully to climate change, many decisions will need to be made far ahead. It is worth bearing in mind, for example, that the Thames Barrier was built in response to the catastrophic east coast floods of 1953 that killed over 300 people in the UK, but it took 30 years to see the project come to fruition. In adapting to a warmer world, decisions that involve major changes to infrastructure, such as coastal developments, capital projects such as new dams, irrigation schemes, airports, or new towns, or those that have potential ramifications for agriculture and biodiversity need to be made up to several decades ahead.

One of the great problems with adaptation is that current forecasts of what we might expect weather- and climate-wise remain rather coarse. This means that they can provide a fair picture at the global and sometimes regional level, but are unable to tell us what to expect at local level. Even Japan's *Earth Simulator*, the world's most powerful supercomputer, only has a resolution of 10 km. Salvation is on the way, however, and you could help. In the world's largest distributed computer project, climatologists are looking for volunteers happy to allow their home computers to run different climate change forecast models when they are not being used for anything else. So, if you want to know what the climate of your street will be in 2054, rush to www.climateprediction.net and download the software now.

Much as we in the UK and the other industrialized countries will have to modify our lives to take account of global warming, the billions of people across the world living in conditions of extreme poverty and deprivation will bear the brunt of climate change. The poorest countries on the planet, who have contributed almost nothing to the rise in GHG emissions, are going to suffer most, perhaps to the extent of costing them each year more than 10 per cent of their incomes. For countries that are already wracked by civil strife, food and water shortages, this added burden will be simply too much to bear. Countries like Bangladesh and some of the Pacific islands, for example, who will see major chunks of their land area return to the sea as water levels rise, will struggle to afford the enormous cost of migration and resettlement. The same goes for the nations of sub-Saharan Africa, for whom the prospects of bigger and longer droughts mean just one thing—more famine and more death. John Schellnhuber, who runs the UK's Tyndall Centre for Climate Change research, has insisted that the industrialized countries who have engineered the global warming catastrophe are duty bound to help out. He suggests that at the very least a UN-supervised adaptation fund of several trillion dollars should be established by 2050, together with a climate insurance scheme within which the premiums are paid for by the industrialized nations via a levy on their emission allowances. Without something along these lines, the world is going to split right down the middle, with the polluting countries using their resources and expertise to adapt and/or muddle through as best they can, and the rest banished to a hellish future defined by starvation, water wars, declining health, falling life expectancies, civil strife, and a breakdown of social and economic order.

Such a terrible prospect for the bulk of the world's population, should the developed world fail to accept its responsibilities, is beginning to turn the heads of some climatologists, including Bert Brolin, former Chairman of the IPCC, towards the need to use technology to stop global warming in its tracks. Even the environmental guru and inventor of the Gaia hypothesis, James Lovelock, is accepting that some quick technological doctoring of the planet might well be needed to bring global warming under control. Seeding the southern ocean with huge quantities of iron filings, like some

giant school chemistry-lab experiment, might have sounded
a little far-fetched, but far wackier and downright mad solutions are currently being dreamt up, by what I earlier referred
to as the 'are you feeling lucky' school, to bring global warming under almost instant control. All involve engineering on
a planetary scale, all are untried, and all carry enormous
risks. So what do they have in store? Maybe I will start with the
most outlandish scheme, unsurprisingly proposed by the late
Edward Teller, 'father' of the H-bomb and allegedly the
model for Dr Strangelove. Before his death in 2003, Teller
and his colleagues at California's Lawrence Livermore
Laboratory came up with the idea of scattering a million
tonnes of tiny (4 mm), hydrogen-filled aluminium balloons
in the stratosphere, the idea being that these would reflect
back sufficient solar radiation to cool the surface. Did I say
this was the craziest idea? I was forgetting about the febrile
imagination of Lowell Wood, co-author of the reflective
balloon paper. In case the balloon idea fails to catch on,
Wood proposes a giant, diaphanous mirror 1,000 km across,
launched bit by bit and parked between our planet and the
Sun. Once again, the goal is to reduce the amount of sunlight reaching the surface, so combating warming. And it
doesn't get any better. John Latham, another US scientist at
the National Center for Atmospheric Research in Boulder,
Colorado, wants to seed clouds over the ocean with salt spray
blasted skywards using giant wind generators. This addition
of millions of tiny salt nuclei will, according to Latham,
increase the number of water droplets in the clouds, making
them whiter and therefore more reflective. A reduction in

solar radiation reaching the surface would again be the result, with lower global temperatures the end product.

The problem I have with these proposals to control our planet's temperature by manipulating its *albedo* or reflectivity is not with the technology, which is certainly feasible, but with the arrogant and cavalier attitude of their proponents. Little, if any, consideration appears to be given to the idea that things might go wrong, that there might be unforeseen side-effects, or that such novel approaches might make the situation worse. Another point that is often missed in discussion of such schemes is that while they may prove effective at reducing surface temperatures, they will have no effect on the level of carbon dioxide and other GHGs in the atmosphere. To engineer the planet so that we can forever tinker with its albedo to balance out the emission of ever more carbon dioxide is to ignore its other effects, such as progressive acidification of the oceans causing reefs and other marine animals with carbonate skeletons to dissolve. But then, what is the extinction of a few thousand species of marine life to a life-form with delusions of omnipotence? The real worry, however, is that the degree of any cooling brought on by giant space mirrors, or a billion reflective mini-balloons, might just turn out to be more than we bargained for. A tiny miscalculation and we could be facing not rising sea levels but falling ones; not vistas of sun-baked earth but the towering fronts of ice sheets escaping their polar fastnesses. And this brings me to one of the most scientifically interesting aspects of global warming. What effect, if any, will it have on the next Ice Age?

We are currently in the middle of an *interglacial* period, a brief respite during which the glaciers have retreated to the poles for a short break before returning with renewed vigour. If the roller-coaster of heat and cold that has characterized the last couple of million years is anything to go by, then temperatures should be on the way down any millennium soon. In the normal run of things, we would expect the world to be a good three degrees C colder in 8,000 years or so, but with temperatures rising at an unprecedented rate will this now happen or could we have postponed the onset of the next Ice Age, perhaps forever? There is some evidence to suggest that this could be the case. No one has yet been able to explain why the most recent batch of Ice Ages—following a number of others in the dim and distant past of the Earth's history—appeared out of the blue around 10 million years ago. One idea, though, is that carbon dioxide levels in the atmosphere, which had been steadily falling from 1,600 ppm around 300 million years ago, dropped below a critical threshold level of maybe 400 ppm. With carbon-dioxide concentrations predicted to be anything between 450 and 1,000 ppm by the end of the century, maybe the ice sheets will have a long time to wait before being released once more from their icy lairs. Then again, it is looking increasingly likely that global warming may bring colder weather—at least to the UK and north-west Europe—and perhaps quite soon.

In the spring of 2004 I was invited to the first UK screening of yet another Hollywood disaster blockbuster, this time a film about climate change called *The Day After Tomorrow*. In the film, rapidly rising Arctic temperatures cause a huge outflow of cold, fresh water into the North Atlantic, bringing the Gulf Stream to a grinding halt and within weeks plunging the world into a new Ice Age. The original premise is fine and the first ten minutes of the film make a reasonable fist of presenting our current knowledge of the relationships between ice melt, ocean currents, and climate. As you might expect, however, it soon goes completely off the rails, with Los Angeles inexplicably obliterated by giant tornadoes after 30 minutes, and the arrival of full Ice Age conditions an hour after that. Nature never moves quickly enough for Hollywood, so everything has to be telescoped down. Even so, we may have only decades to wait before things get decidedly chilly in the UK and north-western Europe, for just those reasons presented in the film.

More and more evidence has accumulated in the last couple of years for ongoing dramatic changes occurring in the circulation of the North Atlantic, and concern is growing that the Gulf Stream, which bathes the UK, Ireland, and Norway in warm waters from the tropics, may already be waning. For a nation in which hydrocarbon supplies are falling and where the transport system grinds to a halt at the sight of a single snowflake, the prospect of ice floes in the North Sea and snow on the ground for 100 days a year is a particularly scary prospect. The Gulf Stream keeps the UK's temperatures up to 10 degrees C higher than they should be, given our latitude, and any significant slow-down could see temperatures fall by 5 degrees C or so, and bring a return to the fierce winters of the Little Ice Age that reigned between the seventeenth and nineteenth centuries.

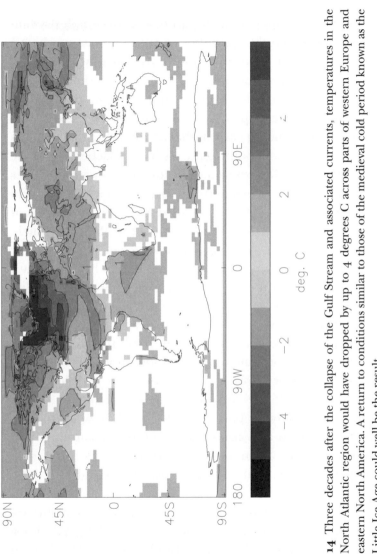

14 Three decades after the collapse of the Gulf Stream and associated currents, temperatures in the North Atlantic region would have dropped by up to 4 degrees C across parts of western Europe and eastern North America. A return to conditions similar to those of the medieval cold period known as the Little Ice Age could well be the result.

The Gulf Stream is actually part of a more complex system of currents known by a number of different names, which I will reduce—for convenience—to the Atlantic Conveyor. This incorporates not only the Gulf Stream but also the counterbalancing cold, deep, return currents that convey water back to the tropics, and the *Sub-polar Gyre*, the broad, anticlockwise-rotating current that carries Gulf Stream waters first west and then south again. As it approaches the Arctic the Gulf Stream loses heat and heads back to warmer climes along the coast of Greenland and eastern Canada in the form of the cold, iceberg-laden current responsible for the loss of the *Titanic*. This is the western segment of the gyre. Much, however, overturns—cooling and sinking beneath the Nordic seas between Norway and Greenland, before heading south again deep below the surface. The worry is that if increased precipitation and ice melting in the Arctic and an increasing northward flow of fresh water from Siberian rivers reduces the salinity of the Gulf Stream, the resulting fall in density will prevent it from sinking. This would ultimately hinder or prevent the return flow of cold, dense water and slow down or even shut off the Atlantic Conveyor circulation. For a while climate models have predicted a future weakening of the Gulf Stream as a consequence of global warming, perhaps a century or more down the line. The latest data suggest, however, that the whole process may already have started.

Bob Dickson, of the Centre for Environment, Fisheries, and Aquaculture Science at Lowestoft in the UK, and his co-workers have reported a sustained and widespread freshening of returning deep waters between Greenland and Norway, which appears to have been going on for the past three or four decades. Already the freshening is extending along the North American eastern seaboard towards the equator. At the same time, Bogi Hansen at the Faroese Fisheries Laboratory, and colleagues in Scotland and Norway, have been monitoring the deep outflow of cold water from the Nordic seas as it passes over the submarine Greenland–Scotland ridge that straddles the North Atlantic at this point. Their results show that the outflow has fallen by 20 per cent since 1950, implying a comparable reduced inflow from the Gulf Stream. It also seems that not only the intensity of the outflow is changing, with new evidence from satellite observations showing that the Sub-polar Gyre has also slowed dramatically throughout the 1990s. So far, there has been no direct indication that the Gulf Stream itself is slowing, but some inkling that this might have started has come from an unexpected source. Although England's lakes and rivers now contain more fish species in greater numbers than ever before, for some reason the eel population is suffering an almighty crash—to just 1 per cent of its former numbers. Every year, eels leave the UK *en masse* and head across the Atlantic to the Sargasso Sea, where they breed. The tiny elvers then drift back on the Gulf Stream to the UK, where they grow and mature before they in turn head to the Sargasso to breed—the entire cycle taking something like 15 years. But now the elvers don't seem to be making it back, and weakening of the Gulf Stream is being blamed as a possible cause, with a slower current preventing the vulnerable elvers from surviving long enough to make it to Blighty.

It is impossible to say when we will lose the warming effects of the Gulf Stream, but the change when it comes is likely to be very rapid, possibly taking just five to ten years. One climatologist puts the chances at 50 : 50 over the next century; another suggests that it could happen any time after 2010. Deep concern in the UK is expressed by the fact that the Natural Environment Research Council recently launched its 20 million-pound RAPID programme, which will address the broad issue of rapid climate change, but will focus in particular on the role of the Atlantic Conveyor. The effects of the loss of Gulf Stream warming on the UK would be devastating, requiring a complete rethink across a range of policies, including energy, transport, and health. Much of northwestern Europe is also likely to suffer from a much colder climate, although the probable degree and extent of the change are not known.

This business with the Gulf Stream teaches us a couple of lessons that we would do well to remember if, as a race, we want to continue to prosper. First, climate does not always change slowly and incrementally. Sometimes it jumps from one state to another in a manner analogous to flipping a light switch rather than turning a dimmer. Examples of such *abrupt climate change* abounded during the Ice Ages, and just 11,000 years ago or so, the rapidly retreating ice sheets were halted in their tracks as plummeting temperatures heralded a 1,000 year freeze known as the *Younger Dryas*. The recent dramatic changes in North Atlantic waters may well provide our first experience of such an event in the flesh. Second, the fact that we can jump from the fire to the fridge so suddenly should remind us that climate-wise we are currently walking a tightrope between hothouse Earth and icehouse Earth. Whatever the ultimate outcome of the great global warming experiment, the fact is that we are still in an interglacial and the next Ice Age still lies in wait several thousand years down the line. Furthermore, should one or other of the aforementioned albedo-changing experiments be tried and go haywire, then the ice sheets could be glistening on the horizon far sooner than this.

You don't have to be a rocket scientist to appreciate the fragility and capriciousness of our planet's climate or to understand that human activities over the last few centuries have proved to be very effective at knocking it out of kilter. You also don't have to be a brain surgeon to realize that while global warming is already here and will impinge upon every last one of us, we have the tools to avoid the worst and to stabilize the climate system before it is too late. We don't need unproven, quick-fix, technological solutions that stand a good chance of making the situation worse, but we do need a coherent and effective plan supported by the *entire* international community to cut back on the rate at which greenhouse gases are accumulating in the atmosphere and ultimately to start bringing the levels down. The Contraction and Convergence model seems best placed to fit the bill, in terms of stabilizing emissions. Furthermore it is fair and attracting growing support from the industrialized countries and the developing world alike. Cutting emissions through

15 A giant mirror in orbit has been proposed to reflect some of the Sun's rays back into space and attempt to bring rising global temperatures under control.

adoption of renewable energy sources and the phasing out of fossil fuels will probably not be sufficient to reduce GHG levels as quickly as we would wish, so some form of sequestration may well have to be adopted. Notwithstanding these measures we are still going to face—in the next hundred years—a global average temperature rise at least double that which has occurred over the last century. Adaptation is therefore a necessity if we are to cope as best we can with a warmer world. Given the political will, a time may come in a century or so when GHGs in the atmosphere will be falling. Then, however, we will be faced with a difficult dilemma? How far do we wish them to fall? Perhaps back to pre-industrial levels, with carbon dioxide concentrations of 280 ppm or maybe not quite so low. Thinking about that tightrope again and the fact that the ice sheets are due to be on the move soon, maybe we should keep levels that little bit higher—at around the 400 ppm mark. The decision, in the end, will not be for us to make, but for the sake of our descendants and the survival of our society, let's hope it is the right one.

Epilogue: Doom or Bloom?

Predicting is very difficult, especially about the future.

Niels Bohr (1885–1962).

I t's time to grow up. In global warming we have created a monster that can only be vanquished by mature and rational thinking alongside cooperation involving all who share the planet and its resources. The time for juvenile bullying, bickering, and point-scoring has gone, and the time for adult negotiation and concerted action has arrived. That action will be taken, I have no doubt; doing nothing is no longer an option. But what will we do? It is no surprise that the arrival of the first gee-gee to impinge upon our civilization—anthropogenic climate change—coincides with our ability to undertake large-scale planetary engineering, because the staggering advances in science and technology over the past three centuries have led to both. This situation presents us with a critical decision, however, placing us at a fork in the road and requiring us to choose which path to follow to our future. Will we take the route that sees us working with the Earth or one that sees us battling it? Is it our choice to seek the wardenship of our planet, protecting and cultivating a world of stunning landscape and wonderful biodiversity, or do we covet the role of despot; harassing and manipulating the earth to meet our own selfish requirements, or at least those of the most powerful? In other words, do we embrace GHG reductions combined with a more sustainable lifestyle as the solution for global warming, or do we go for the business as usual/planetary engineering solution? Which we choose could affect the lives and deaths of billions stretching many centuries ahead.

The two options are not mutually exclusive, in the sense that both are hugely dependent upon the most cutting-edge of technologies. Rather, they reflect different points of view about what progress means. On the one hand, a sustainable journey measured by an improved quality of life for all, on the other an ultimately unsustainable voyage defined purely by the pace and scale of economic growth. Personally, I would plump every time for the former, and I hope very much that this will be the path we follow. I have to admit that I have never quite understood why everything has to get bigger every year, from McDonald's burgers to the people who eat them. Why, every January, does it seem to be a catastrophe if Christmas sales were not once again bigger than ever before? Big is clearly not beautiful, as a 2002 Cabinet Office paper demonstrates. Survey results published in the soberly titled *Life Satisfaction: The State of Knowledge and Implications for Government* revealed that in the previous three decades, average individual prosperity in the UK had risen 80 per cent while the Life Satisfaction Index had stayed pretty much the same. So richer does not mean happier—as an individual, as a family, as a nation, or as a civilization—but then, we all knew that, didn't we? Perhaps it is worth reminding ourselves of the words of the great Ernst Schumacher, author of *Small is Beautiful* way back in 1973, but just as relevant in today's world of climate change and globalization.

> Ever bigger machines, entailing ever bigger concentrations of power and exerting ever greater violence against the environment, do not represent progress: they are a denial of wisdom. Wisdom demands a new orientation of science and technology towards the organic, the gentle, the non-violent, the elegant and beautiful.

If we choose the warden's path to tackling global warming, we will have no alternative but to lead a more sustainable lifestyle, along the lines Schumacher espoused, with the onus on personal happiness, an improved environment, and a fairer world. If we follow in the footsteps of the despot then growth will continue unabated and at all costs. But this second path has other pitfalls too. The quick-fix, albedo-modifying solutions for global warming are untried and untested. Failure is not an option, yet there is no guarantee of success. Their proponents seem utterly oblivious of the *precautionary principle*, or perhaps they are just dismissive of its importance. There are a number of definitions of the principle, some wordier and more convoluted than others, but the award-winning science writer, Colin Tudge, puts it simply thus: 'Of course you can make no progress without risk. But if there is no obvious gain from taking the risk, then don't take it.' If scientists had not taken risks in the past then we would be in a sorry state: no aeroplanes, no cars, no antibiotics, no transplants; then again, no H-bomb, no nerve gas, no bio-weapons. While development of the former would have passed the precautionary principle test with flying colours, there is little doubt that the latter—due to their potential for triggering planet wide catastrophe—would have bitten the dust. When dealing with the fate of the one and only planet we have, and particularly when another perfectly reasonable course of action is available to us, there can be no excuse for allowing the sort of mega-scale meddling proposed by the late Edward Teller and his cronies. Engineering the planet to solve a problem that has arisen ultimately as a result of the misuse of science and technology provides too good an opportunity for the see-saw effect to come into play. To reiterate, this involves using science and technology to solve a problem that they created in the first place, but making the situation worse. Then trying to tackle the new situation but making things even more dire, and so on. The precautionary principle provides us with a vital safeguard to prevent just such a sequence of events, so let's use it.

While I am confident that the warden's path is the one we should be following, I don't have a say, and nor do most of you. In fact, as I discussed in Chapter 1, in relation to the impact threat, no international protocol currently exists that defines how momentous decisions relating to the welfare of the planet and the race should be taken and who should take them. Worryingly, however, there are some signs that when the IPCC meets again in a couple of years, at least some representatives may be prepared to consider contenders for

the quick-fix, albedo-modifying approach to accelerating warming. Any move to seek an international consensus supporting this approach must be fought tooth and nail, not only because of the great risks involved but also because, at a stroke, we will have changed utterly our relationship with the planet, perhaps permanently and almost certainly to our detriment and that of all life that shares our world.

Whichever route we decide to follow in tackling global warming, and whatever the outcome, we will still—at some point in the future—have to face a gee-gee of a different sort. An asteroid or comet on a potential collision course will be discovered, magma will breach the surface in a cataclysmic super-eruption, and La Palma's Cumbre Vieja will crash into the North Atlantic. I hope that in the preceding chapters I have been able to convince you that—as with climate change—we have the tools to cope: if not to prevent, then at least to mitigate the worst effects and manage the aftermath and recovery. I would like to think that following the warden's path to a cooler planet would result in a closer-knit global community that is better placed successfully to take on another gee-gee when the need arises, using science and technology prudently and with discretion to the benefit of all.

And so to the big question? Are we doomed or will we bloom? Barring the impact of a giant comet or asteroid, none of the gee-gees I have introduced are capable of wiping out our race, so some—probably many—of us *will* survive Armageddon, whatever form it takes. Of course, Martin Rees may turn out to be right, and we could all be rendered down into a mass of grey nano-goo before the century is out, but then at least be wouldn't have global warming to worry about. In the epilogue to *A Guide to the End of the World: Everything You Never Wanted to Know,* I painted a none too rosy picture of our planet a century hence: a hotter and more hazardous world, half as populous again as it is now and dominated by an increasingly elderly age-group. A world in which the sixth great mass extinction had already decimated other living things, wiping out maybe a quarter or more of all land animals and plants and massively reducing biodiversity for millions of years to come. To be honest, this perspective hasn't changed, and the picture painted is likely to be pretty accurate. The point is, however, that things could be far worse, and everything is still very much up for grabs. If, by AD 2100, we have stabilized GHG emissions at 450 ppm or less without recourse to planetary engineering, then we would have had to have successfully made the jump to a more sustainable and cooperative world within which far greater value is attached to the welfare of the environment and future generations. Should GHG concentrations in the atmosphere be hovering about the 800 ppm mark by the end of the century, then hothouse Earth will truly have arrived, with conditions far more desperate even than those imagined in my last book. Of course, should the planetary engineers be given free rein to cock things up, we could by then be scavenging for food scraps along the ice-bound streets of central London or New York.

Now, whenever I give a public lecture, I have taken to describing myself as an optimistic pessimist, mainly to try to shake off the doom-monger label that follows me around like a bad smell. What I really mean by this, however, is that I am fully aware that things are going to get worse, but in the long run I am hopeful that we—as a race—have everything we need to pull through and ultimately to prosper. The global warming challenge is providing us with the perfect opportunity to pull the world's nations together into a community with a more mature outlook capable of looking ahead—not just to the next election five years down the line—but into the next century and beyond. This is the gee-gee perspective, and it is *exactly* the outlook that is required if we are to protect ourselves and our planet from potentially devastating geophysical events that only happen very infrequently. If we can grasp the opportunities presented by the climate-change crisis and by the dreadful Indian Ocean tsunami, to allow us to see the bigger picture and plan for the longer term, then our race, our planet, and all life upon it can bloom. If we are truly to survive Armageddon, then we mustn't let this unique chance slip.

Further Reading

Bryant, Edward. *Tsunami: The Underrated Hazard.* Cambridge University Press, 2001.

Hadfield, Peter. *Sixty Seconds that will Change the World: The Coming Tokyo Earthquake.* Rev. edn. Pan, 1995.

Hillman, Mayer. *How we can Save the Planet.* Penguin, 2004.

Houghton, John. *Global Warming: The Complete Briefing.* Cambridge University Press, 2004.

Intergovernmental Panel on Climate Change (IPCC) Working Group II. *Climate Change 2001: Impacts, Adaptation and Vulnerability.* Cambridge University Press, 2001.

Leggett, Jeremy. *The Carbon War.* Penguin, 2001.

Lewis, John. *Comet and Asteroid Impact Hazards on a Populated Earth.* Academic Press, 1999.

Lynas, Mark. *High Tide: News from a Warming World.* Flamingo, 2004.

McGuire, Bill *Apocalypse: A Natural History of Global Disasters.* Cassell, 1999.

McGuire, Bill. *Raging Planet: Earthquakes, Volcanoes and the Tectonic Threat to Life on Earth.* Apple Press, 2002.

McGuire, Bill. *A Guide to the End of the World: Everything You Never Wanted to Know.* Oxford University Press, 2002.

McGuire, Bill, Mason, Ian, and Kilburn, Christopher. *Natural Hazards and Environmental Change.* Hodder Arnold, 2002.

Meyer, Aubrey. *Contraction & Convergence: The Global Solution to Climate Change.* Green Books, 2000.

Mulhall, Douglas. *Our Molecular Future.* Prometheus Books, 2002.

Steel, Duncan. *Target Earth.* Reader's Digest, 2000.

Index

BP 179
Brazil 19, 135
Brewer, Peter 158–9
British Association Festival of
    Science (2003) 89
Broeker, Wallace 158
Brolin, Bert 196
Bronze Age 76
Bryant, Nevin 149
building
    codes/design/construction
    31, 152–4
bullet trains 152
Bush, George W. 7, 84, 97, 173

calderas 40, 116–17, 120
California 37, 72, 85, 99–100,
    101, 148
  central 152
  earthquake threat 140
  Lawrence Livermore
    Laboratory 197
  longest rupture ever recorded
    141
  southern 150, 152
  see also Los Angeles; Monterey;
    San Francisco
California Earthquake Prediction
    Evaluation Council 150
Campbell, Jonathan 95
Campi Flegrei 120, 122
Canada 52, 53, 54, 170, 186, 202
Canary Islands 27, 129, 137
  see also El Hierro;
    Fuerteventura; Gran
    Canaria; La Palma; Tenerife

C&C (Contraction and
    Convergence) model 171,
    172, 205
Cape Verde Islands 25, 139
CAPS (NASA Comet/Asteroid
    Protection System) 73, 95,
    96, 97
carbon dioxide 39, 51, 54, 125,
    159–73, 188–90
  drop below critical threshold
    level 199
  rationed emissions 191
Carbon Dioxide Capture and
    Storage scheme (UK) 162
carbon unit plan 191–2
Caribbean 19, 23, 134, 135
  warning systems 137
  see also Martinique; Montserrat
Carracedo, Juan Carlos 12
Central America 115
Central Asia 174
Centre for Environment,
    Fisheries, and Aquaculture
    Science (UK) 202
Ceres (asteroid) 80
Cerro Hudson 34
chemical rockets 91
Chicxulub impact 49
Chile 34
China 54, 57, 88, 149
  air-conditioning 187
  forest clearances 169
  free hand to increase GHG
    emissions 170
  massive reforestation 165
chokka-gata quakes 143